Islet Inflammation and Metabolic Homeostasis

Islet Inflammation and Metabolic Homeostasis

Editors

Jason Collier
Susan Burke

MDPI • Basel • Beijing • Wuhan • Barcelona • Belgrade • Manchester • Tokyo • Cluj • Tianjin

Editors
Jason Collier
Pennington Biomedical
Research Center
USA

Susan Burke
Pennington Biomedical
Research Center
USA

Editorial Office
MDPI
St. Alban-Anlage 66
4052 Basel, Switzerland

This is a reprint of articles from the Special Issue published online in the open access journal *Metabolites* (ISSN 2218-1989) (available at: https://www.mdpi.com/si/metabolites/Islet_Inflammation).

For citation purposes, cite each article independently as indicated on the article page online and as indicated below:

LastName, A.A.; LastName, B.B.; LastName, C.C. Article Title. *Journal Name* **Year**, *Volume Number*, Page Range.

ISBN 978-3-0365-0926-6 (Hbk)
ISBN 978-3-0365-0927-3 (PDF)

Contents

About the Editors

Jason Collier studies islet biology at the Pennington Biomedical Research Center in Baton Rouge, Louisiana. His laboratory focuses on cytokine signaling in beta cells using in vitro and in vivo approaches.

Susan Burke studies the role of lipid metabolism in islet beta-cell function using rodent models of obesity and diabetes. Her research is conducted at Pennington Biomedical Research Center in Baton Rouge, Louisiana.

metabolites

Special Issue: Islet Inflammation and Metabolic Homeostasis

Susan J. Burke [1,*] and J. Jason Collier [2,3,*]

1 Laboratory of Immunogenetics, Pennington Biomedical Research Center, Baton Rouge, LA 70808, USA
2 Laboratory of Islet Biology and Inflammation, Pennington Biomedical Research Center, Baton Rouge, LA 70808, USA
3 Department of Biological Sciences, Louisiana State University, Baton Rouge, LA 70803, USA
* Correspondence: Susan.burke@pbrc.edu (S.J.B.); Jason.collier@pbrc.edu (J.J.C.); Tel.: +1-225-763-2532 (S.J.B.); +1-225-763-2884 (J.J.C.)

Citation: Burke, S.J.; Collier, J.J. Special Issue: Islet Inflammation and Metabolic Homeostasis. *Metabolites* **2021**, *11*, 77. https://doi.org/10.3390/metabo11020077

Received: 22 January 2021
Accepted: 26 January 2021
Published: 28 January 2021

Publisher's Note: MDPI stays neutral with regard to jurisdictional claims in published maps and institutional affiliations.

This special issue was commissioned to offer a source of distinct viewpoints and novel data that capture some of the subtleties of the pancreatic islet, especially in relation to adaptive changes that influence metabolic homeostasis. As such, this forum provided a home for review articles that offered original perspectives relevant to islet inflammation, tissue resident immune cells, and related variables that influence these factors. In addition, we are pleased to note that articles in this issue also included primary data addressing rapidly evolving and exciting areas of research. Indeed, the pancreatic islet is a magnificent micro-organ that is capable of rapid adaptive responses to support the changing needs of the organism [1].

A very timely study by Piñeros et al. shows the impact of one week of high-fat feeding on changes at the single-cell level in mouse islets [2]. This is important for two reasons: (1) after several months on a high-fat diet, islet adaptations (e.g., increases in insulin-positive cell mass, etc.) have already taken place [3,4]. Thus, investigating early time points is essential to understand the initial alterations required for adaptive responses. (2) Outcomes associated with increases in beta-cell mass, such as enhanced proliferation markers, arise early (within days) after a stimulus that promotes insulin resistance (e.g., high-fat feeding, glucocorticoid exposure, etc.) [4–6]. Consequently, an analysis of individual cell populations and how these cells cluster into groups one week after high-fat feeding is important complementary information that offers a window into the individual cell level during this adaptive response. The authors found that not all islet β-cells responded the same to one week of high-fat feeding. Indeed, the data were binned into three major clusters and seven minor clusters. The greatest differences were identified within the minor clusters. This fits with overall β-cell heterogeneity and provides unique insight into early β-cell responses prior to the development of overt glucose intolerance and measurable insulin resistance.

Discovering the signaling pathways that promote an increase in β-cell mass as well as those that enhance insulin secretion reflect ongoing research efforts. How macrophages contribute to such alterations in the islet is an important consideration for adaptive responses to insulin resistance and perhaps also to autoimmunity. A review by Jensen and colleagues offers insights into the dual role that macrophages may have in islet biology [7]. Macrophages are derived from circulating monocytes and also arise through conditioning in the tissue where they ultimately reside [8]. This underscores an important role for macrophages to not only support the growth of β-cells developmentally [9,10], but also the probability that they support the growth and function of mature adult β-cells during adaptive responses [11,12].

Locatelli and Mulvihill put forth a viewpoint on the role of low carbohydrate (<30% kcal) or ketogenic (<10% kcal) diets on three specific parameters: islet health with a focus on the preservation of beta-cell mass in models of rodent obesity and diabetes, secretion of islet hormones to maintain glucose homeostasis, and the response of peripheral tissues to insulin. The effects of such diets in pre-clinical studies and clinical outcomes were

1

carefully considered [13]. Furthermore, the translational relevance of low carbohydrate and ketogenic diets is clear when comparing the impact of reducing carbohydrates in settings of T1D and T2D [13,14].

With the knowledge that diets are linked to inflammation and islet β-cell dysfunction, several reviews in this Special Issue focus on stimuli that promote islet β-cell inflammation as well as the outcomes associated with oxidative stress and ER stress. Two reviews by M. Cerf provide complementary information on the glucose and lipid toxicity (sometimes referred to collectively as glucolipotoxicity) contribution to dysfunction in β-cells [15,16]. A third review provides an in-depth perspective on oxidative stress from the viewpoint of the NADPH oxidase superfamily [17]. This is particularly important considering how immune cell-derived cytokines have been shown to promote nitrosative and oxidative stress in β-cells [18,19] and the possible linkage between Nox2 and nutrient stress [20]. Moreover, whether glucose is the major culprit, as opposed to fatty acids, is also relevant and raises important questions for the field as a whole to consider [21]. The reviews in this Special Issue offer additional perspectives towards that discussion. Indeed, with β-cells having a high metabolic rate and being a major site of glucose metabolism [22], a deeper understanding of the role of oxidative, nitrosative, inflammation-based, and ER stress is essential to the goal of enhancing proliferation for therapeutic purposes as well as preventing losses in insulin secretion that lead to diabetes.

When contemplating the role of fatty acids in diabetes and islet dysfunction, it should be noted that such lipids come in many forms, including differing degrees of saturation as well as chain lengths (i.e., short, medium, long, etc.). The role of lipids is complex, with discrete species viewed as pro-inflammatory while others have been attributed as having anti-inflammatory properties [23–25]. An interesting contribution by the Kimple laboratory explores fatty acids that activate the EP3 receptor, an important G-protein coupled receptor that responds to the arachidonic acid metabolite prostaglandin E_2 (PGE_2) as its primary ligand. In this study, the authors used the Black and Tan BRachyury (BTBR) mice homozygous for the *ob/ob* mutation (leptin-deficient) and found there was a differential effect of diet, when combined with $BTBR^{ob}$ genotype, that impacted the onset of hyperglycemia [26]. In addition, the authors found that changes in gut microbiota were linked with hyperglycemia versus normoglycemia. Moreover, an altered interleukin-1 and PGE^2/EP3 signaling link with changes in islet β-cells in this mouse model was observed.

Taken together, the articles in this issue provide unique insights into islet inflammation and metabolic homeostasis. As guest editors, we are grateful for the quality of articles and the discrete and complementary nature of the work contributed to this Special Issue. We look forward to the continued advancement of knowledge in the field of islet biology that builds from the studies presented here. We want to thank the peer reviewers who provided the rigorous scientific evaluation of each submission as well as the members of the Metabolites Editorial Office for their support during the development, review process, and final steps needed to complete this issue of Metabolites.

Author Contributions: Conceptualization, S.J.B. and J.J.C.; writing—original draft preparation, S.J.B. and J.J.C.; writing—review and editing, S.J.B. and J.J.C.; funding acquisition, S.J.B. and J.J.C. All authors have read and agreed to the published version of the manuscript.

Funding: Studies in the Collier laboratory are supported by NIH grants R21 AI138136, R01 DK123183, R01 DK108765, and R03 AI151920. Pilot and feasibility funding via NIH P30 DK072476 as well as NIH grants P20 GM135002 and R01 DK108765 support research in the Burke laboratory.

Conflicts of Interest: The authors declare no conflict of interest. The funders had no role in the writing of or the decision to publish this editorial.

References

1. Burke, S.J.; Karlstad, M.D.; Collier, J.J. Pancreatic Islet Responses to Metabolic Trauma. *Shock* **2016**, *46*, 230–238. [CrossRef]
2. Pineros, A.R.; Gao, H.; Wu, W.; Liu, Y.; Tersey, S.A.; Mirmira, R.G. Single-Cell Transcriptional Profiling of Mouse Islets Following Short-Term Obesogenic Dietary Intervention. *Metabolites* **2020**, *10*, 513. [CrossRef]

3. Burke, S.J.; Batdorf, H.M.; Burk, D.H.; Noland, R.C.; Eder, A.E.; Boulos, M.S.; Karlstad, M.D.; Collier, J.J. db/db Mice Exhibit Features of Human Type 2 Diabetes That Are Not Present in Weight-Matched C57BL/6J Mice Fed a Western Diet. *J. Diabetes Res.* **2017**, *2017*, 8503754. [CrossRef]

4. Mosser, R.E.; Maulis, M.F.; Moulle, V.S.; Dunn, J.C.; Carboneau, B.A.; Arasi, K.; Pappan, K.; Poitout, V.; Gannon, M. High-fat diet-induced beta-cell proliferation occurs prior to insulin resistance in C57Bl/6J male mice. *Am. J. Physiol. Endocrinol. Metab.* **2015**, *308*, E573–E582. [CrossRef]

5. Stamateris, R.E.; Sharma, R.B.; Hollern, D.A.; Alonso, L.C. Adaptive beta-cell proliferation increases early in high-fat feeding in mice, concurrent with metabolic changes, with induction of islet cyclin D2 expression. *Am. J. Physiol. Endocrinol. Metab.* **2013**, *305*, E149–E159. [CrossRef]

6. Burke, S.J.; Batdorf, H.M.; Huang, T.Y.; Jackson, J.W.; Jones, K.A.; Martin, T.M.; Rohli, K.E.; Karlstad, M.D.; Sparer, T.E.; Burk, D.H.; et al. One week of continuous corticosterone exposure impairs hepatic metabolic flexibility, promotes islet beta-cell proliferation, and reduces physical activity in male C57BL/6J mice. *J. Steroid Biochem. Mol. Biol.* **2019**, *195*, 105468. [CrossRef] [PubMed]

7. Jensen, D.M.; Hendricks, K.V.; Mason, A.T.; Tessem, J.S. Good Cop, Bad Cop: The Opposing Effects of Macrophage Activation State on Maintaining or Damaging Functional beta-Cell Mass. *Metabolites* **2020**, *10*, 485. [CrossRef]

8. Calderon, B.; Carrero, J.A.; Ferris, S.T.; Sojka, D.K.; Moore, L.; Epelman, S.; Murphy, K.M.; Yokoyama, W.M.; Randolph, G.J.; Unanue, E.R. The pancreas anatomy conditions the origin and properties of resident macrophages. *J. Exp. Med.* **2015**, *212*, 1497–1512. [CrossRef] [PubMed]

9. Geutskens, S.B.; Otonkoski, T.; Pulkkinen, M.A.; Drexhage, H.A.; Leenen, P.J. Macrophages in the murine pancreas and their involvement in fetal endocrine development in vitro. *J. Leukoc. Biol.* **2005**, *78*, 845–852. [CrossRef] [PubMed]

10. Banaei-Bouchareb, L.; Gouon-Evans, V.; Samara-Boustani, D.; Castellotti, M.C.; Czernichow, P.; Pollard, J.W.; Polak, M. Insulin cell mass is altered in Csf1op/Csf1op macrophage-deficient mice. *J. Leukoc. Biol.* **2004**, *76*, 359–367. [CrossRef]

11. Burke, S.J.; Batdorf, H.M.; Burk, D.H.; Martin, T.M.; Mendoza, T.; Stadler, K.; Alami, W.; Karlstad, M.D.; Robson, M.J.; Blakely, R.D.; et al. Pancreatic deletion of the interleukin-1 receptor disrupts whole body glucose homeostasis and promotes islet beta-cell de-differentiation. *Mol. Metab.* **2018**, *14*, 95–107. [CrossRef] [PubMed]

12. Dror, E.; Dalmas, E.; Meier, D.T.; Wueest, S.; Thevenet, J.; Thienel, C.; Timper, K.; Nordmann, T.M.; Traub, S.; Schulze, F.; et al. Postprandial macrophage-derived IL-1beta stimulates insulin, and both synergistically promote glucose disposal and inflammation. *Nat. Immunol.* **2017**, *18*, 283–292. [CrossRef] [PubMed]

13. Locatelli, C.A.A.; Mulvihill, E.E. Islet Health, Hormone Secretion, and Insulin Responsivity with Low-Carbohydrate Feeding in Diabetes. *Metabolites* **2020**, *10*, 455. [CrossRef]

14. Lennerz, B.S.; Barton, A.; Bernstein, R.K.; Dikeman, R.D.; Diulus, C.; Hallberg, S.; Rhodes, E.T.; Ebbeling, C.B.; Westman, E.C.; Yancy, W.S., Jr.; et al. Management of Type 1 Diabetes With a Very Low-Carbohydrate Diet. *Pediatrics* **2018**, *141*. [CrossRef] [PubMed]

15. Cerf, M.E. Developmental Programming and Glucolipotoxicity: Insights on Beta Cell Inflammation and Diabetes. *Metabolites* **2020**, *10*, 444. [CrossRef]

16. Cerf, M.E. Beta Cell Physiological Dynamics and Dysfunctional Transitions in Response to Islet Inflammation in Obesity and Diabetes. *Metabolites* **2020**, *10*, 452. [CrossRef]

17. Kowluru, A. Oxidative Stress in Cytokine-Induced Dysfunction of the Pancreatic Beta Cell: Known Knowns and Known Unknowns. *Metabolites* **2020**, *10*, 480. [CrossRef]

18. Broniowska, K.A.; Oleson, B.J.; Corbett, J.A. beta-cell responses to nitric oxide. *Vitam. Horm.* **2014**, *95*, 299–322. [CrossRef]

19. Burke, S.J.; Stadler, K.; Lu, D.; Gleason, E.; Han, A.; Donohoe, D.R.; Rogers, R.C.; Hermann, G.E.; Karlstad, M.D.; Collier, J.J. IL-1beta reciprocally regulates chemokine and insulin secretion in pancreatic beta-cells via NF-kappaB. *Am. J. Physiol. Endocrinol. Metab.* **2015**, *309*, E715–E726. [CrossRef]

20. Kowluru, A.; Kowluru, R.A. Phagocyte-like NADPH oxidase [Nox2] in cellular dysfunction in models of glucolipotoxicity and diabetes. *Biochem. Pharmacol.* **2014**, *88*, 275–283. [CrossRef] [PubMed]

21. Weir, G.C. Glucolipotoxicity, beta-Cells, and Diabetes: The Emperor Has No Clothes. *Diabetes* **2020**, *69*, 273–278. [CrossRef] [PubMed]

22. Campbell, J.E.; Newgard, C.B. Mechanisms controlling pancreatic islet cell function in insulin secretion. *Nat. Rev. Mol. Cell Biol.* **2021**. [CrossRef] [PubMed]

23. Serhan, C.N.; Levy, B.D. Resolvins in inflammation: Emergence of the pro-resolving superfamily of mediators. *J. Clin. Investig.* **2018**, *128*, 2657–2669. [CrossRef] [PubMed]

24. Unger, R.H. Lipotoxic diseases. *Annu. Rev. Med.* **2002**, *53*, 319–336. [CrossRef] [PubMed]

25. Fritsche, K.L. The science of fatty acids and inflammation. *Adv. Nutr.* **2015**, *6*, 293S–301S. [CrossRef] [PubMed]

26. Schaid, M.D.; Zhu, Y.; Richardson, N.E.; Patibandla, C.; Ong, I.M.; Fenske, R.J.; Neuman, J.C.; Guthery, E.; Reuter, A.; Sandhu, H.K.; et al. Systemic Metabolic Alterations Correlate with Islet-Level Prostaglandin E2 Production and Signaling Mechanisms That Predict beta-Cell Dysfunction in a Mouse Model of Type 2 Diabetes. *Metabolites* **2021**, *11*, 58. [CrossRef]

 *metabolites*

Article

Systemic Metabolic Alterations Correlate with Islet-Level Prostaglandin E$_2$ Production and Signaling Mechanisms That Predict β-Cell Dysfunction in a Mouse Model of Type 2 Diabetes

Michael D. Schaid [1,2,3], Yanlong Zhu [4,5], Nicole E. Richardson [1,3], Chinmai Patibandla [1,3], Irene M. Ong [6,7], Rachel J. Fenske [2,3], Joshua C. Neuman [2,3], Erin Guthery [1,3], Austin Reuter [1,3], Harpreet K. Sandhu [1,3], Miles H. Fuller [8,9], Elizabeth D. Cox [10], Dawn B. Davis [1,2,3], Brian T. Layden [8,9], Allan R. Brasier [1,11], Dudley W. Lamming [1,2,3], Ying Ge [4,5,12] and Michelle E. Kimple [1,2,3,4,*]

[1] Department of Medicine, University of Wisconsin-Madison, Madison, WI 53705, USA; michael.schaid@northwestern.edu (M.D.S.); nicole.cummings@wisc.edu (N.E.R.); cpatiban@medicine.wisc.edu (C.P.); eringuthery@gmail.com (E.G.); austin.reuter@vumc.org (A.R.); hbrar@wisc.edu (H.K.S.); dbd@medicine.wisc.edu (D.B.D.); abrasier@wisc.edu (A.R.B.); dlamming@medicine.wisc.edu (D.W.L.)
[2] Interdepartmental Graduate Program in Nutritional Sciences, University of Wisconsin-Madison, Madison, WI 53706, USA; rjfenske@wisc.edu (R.J.F.); jnum@novonordisk.com (J.C.N.)
[3] Research Service, William S. Middleton Memorial Veterans Hospital, Madison, WI 53705, USA
[4] Department of Cell and Regenerative Biology, University of Wisconsin-Madison, Madison, WI 53705, USA; yzhu353@wisc.edu (Y.Z.); ying.ge@wisc.edu (Y.G.)
[5] Human Proteomics Program, School of Medicine and Public Health, University of Wisconsin-Madison, Madison, WI 53705, USA
[6] Department of Obstetrics and Gynecology, University of Wisconsin-Madison, Madison, WI 53715, USA; ong@cs.wisc.edu
[7] Department of Biostatistics and Medical Informatics, University of Wisconsin-Madison, Madison, WI 53705, USA
[8] Division of Endocrinology, Diabetes, and Metabolism, University of Illinois at Chicago, Chicago, IL 60612, USA; miles@bcg.com (M.H.F.); blayde1@uic.edu (B.T.L.)
[9] Jesse Brown Veterans Affairs Medical Center, Chicago, IL 60612, USA
[10] Department of Pediatrics, University of Wisconsin-Madison, Madison, WI 53792, USA; ecox@wisc.edu
[11] Institute for Clinical and Translational Research, University of Wisconsin-Madison, Madison, WI 53705, USA
[12] Department of Chemistry, University of Wisconsin-Madison, Madison, WI 53706, USA
* Correspondence: mkimple@medicine.wisc.edu; Tel.: +1-1-608-265-5627

Citation: Schaid, M.D.; Zhu, Y.; Richardson, N.E.; Patibandla, C.; Ong, I.M.; Fenske, R.J.; Neuman, J.C.; Guthery, E.; Reuter, A.; Sandhu, H.K.; et al. Systemic Metabolic Alterations Correlate with Islet-Level Prostaglandin E$_2$ Production and Signaling Mechanisms That Predict β-Cell Dysfunction in a Mouse Model of Type 2 Diabetes. *Metabolites* **2021**, *11*, 58. https://doi.org/10.3390/metabo11010058

Received: 1 November 2020
Accepted: 7 January 2021
Published: 16 January 2021

Publisher's Note: MDPI stays neutral with regard to jurisdictional claims in published maps and institutional affiliations.

Abstract: The transition from β-cell compensation to β-cell failure is not well understood. Previous works by our group and others have demonstrated a role for Prostaglandin EP3 receptor (EP3), encoded by the *Ptger3* gene, in the loss of functional β-cell mass in Type 2 diabetes (T2D). The primary endogenous EP3 ligand is the arachidonic acid metabolite prostaglandin E$_2$ (PGE$_2$). Expression of the pancreatic islet EP3 and PGE$_2$ synthetic enzymes and/or PGE$_2$ excretion itself have all been shown to be upregulated in primary mouse and human islets isolated from animals or human organ donors with established T2D compared to nondiabetic controls. In this study, we took advantage of a rare and fleeting phenotype in which a subset of Black and Tan BRachyury (BTBR) mice homozygous for the *Leptin^{ob/ob}* mutation—a strong genetic model of T2D—were entirely protected from fasting hyperglycemia even with equal obesity and insulin resistance as their hyperglycemic littermates. Utilizing this model, we found numerous alterations in full-body metabolic parameters in T2D-protected mice (e.g., gut microbiome composition, circulating pancreatic and incretin hormones, and markers of systemic inflammation) that correlate with improvements in EP3-mediated β-cell dysfunction.

Keywords: obesity; type 2 diabetes; insulin resistance; inflammation; gut microbiome; untargeted plasma metabolomics; polyunsaturated fatty acids; prostaglandins; insulin secretion; beta-cell function

1. Introduction

Type 2 diabetes (T2D) is characterized by a systemic loss of blood glucose homeostasis that is primarily linked to obesity, which is often associated with insulin resistance (IR) and systemic inflammation. IR and inflammation induce stress on the pancreatic β-cell to meet the demand for enhanced insulin secretion to maintain glucose homeostasis [1]. An inability of β-cells to compensate by increasing mass, functionality, or both results in β-cell failure and hyperglycemia [2].

The transition from β-cell compensation to β-cell failure is not well understood. Our group and others have demonstrated a role for Prostaglandin EP3 receptor (EP3) (gene symbol: *Ptger3*), a G protein-coupled receptor (GPCR) for the arachidonic acid metabolite PGE_2, as a significant contributor to β-cell dysfunction and loss of functional β-cell mass in T2D [3,4]. Most of our prior understanding of the islet PGE_2/EP3 signaling pathway and its role in T2D pathophysiology utilized pancreatic islets isolated from human organ donors with T2D and mouse models with very elevated blood glucose levels compared to their lean, nondiabetic counterparts. One of the most reproducible T2D models utilizes the $Leptin^{ob/ob}$ mutation in the Black and Tan BRachyury (BTBR) mouse strain ($BTBR^{ob}$) [3,4]. In these mice, islet *Ptger3* expression and PGE_2 synthesis are dramatically upregulated, consequently suppressing insulin secretion [3,4]. Treating $BTBR^{ob}$ islets with a specific EP3 antagonist or feeding $BTBR^{ob}$ mice a diet low in arachidonic acid improves their glucose-stimulated insulin secretion (GSIS) response [3,4]. However, PGE_2 has also been linked with beneficial effects on the islet phenotype. For example, PGE_2 promotes M2 macrophage polarization, preventing pro-inflammatory cytokine production and promoting β-cell survival [5]. Therefore, what physiological changes might transition islet EP3 signaling from protective to detrimental remain unknown.

It is well-known that $BTBR^{ob}$ mice of both sexes rapidly and reproducibly become hyperglycemic because of underlying defects in both beta-cell function and skeletal muscle insulin sensitivity [6]. By 10 weeks of age, male $BTBR^{ob}$ mice have end-stage diabetes, with a mean blood glucose of approximately 600 mg/dL [6]. Ten-week-old female mice have mean blood glucose levels of approximately 450 mg/dL and, by 14 weeks of age, have also progressed to end-stage diabetes [6]. The beta-cell-centric nature of BTBROb diabetes progression combined with the rapid and reproducible nature of the phenotype makes this line ideal for our studies. In the process of breeding $BTBR^{ob}$ mice for downstream analyses, we discovered, for an approximately 6 month period, that no 10-week-old $BTBR^{ob}$ mice in our investigator-accessible facility were hyperglycemic. This phenotype was partially related to changes in circulating metabolites after the brand of standard rodent chow was switched (explored in a different work [7]) but was not fully related to diet composition, as ultimately, the phenotype disappeared. Our previous publication found that the biggest differences in circulating metabolites in $BTBR^{Ob}$ mice was phenotype and not diet composition and that these alterations correlated directly with beta-cell function [7]. Therefore, in this work, mice were grouped by phenotype, independent of diet. Using a discovery-based approach, we found that several metabolism-related phenotypes, including gut microbiome composition, levels of circulating pancreatic and incretin hormones, and inflammation-associated adipokines, were dramatically normalized in normoglycemic $BTBR^{ob}$ mice compared to their T2D littermates. High-throughput untargeted metabolomics identified changes in circulating fatty acid conjugates associated with alterations in islet plasma membrane fatty acid composition downstream of PGE_2 synthesis in plasma from T2D vs. normoglycemic $BTBR^{ob}$ mice. Combined with a significant upregulation of islet *Ptger3* expression and the impact of an EP3-selective agonist on GSIS, our results link full-body metabolic derangements specifically with the EP3-mediated β-cell dysfunction of T2D.

2. Results

2.1. Initial Observation That Metabolic Phenotype Can Be Indpendent of Genotype in the BTBR^{ob} Mouse Model of T2D

In our work, we defined a mouse as hyperglycemic when its random-fed blood glucose level was ≥300 mg/dL, as measured using a veterinary glucometer and rat/mouse test strips: a level that should be well-exceeded by both male and female BTBR^{Ob} mice by 10 weeks of age. Unless humanely euthanized, BTBR^{Ob} mice succumb to diabetic ketoacidosis well before 20 weeks of age. We discovered a time-limited phenotype in which the random-fed blood glucose levels of our experimental BTBR^{Ob} mice did not consistently exceed 300 mg/dL until more than 20 weeks of age, with some mice living until 30 weeks of age before end-stage hyperglycemia was apparent (Figure 1). As our experimental mice were transferred from a breeding core facility to an investigator-accessible facility and other facilities across campus did not show the same phenotype (J.C.N. and M.E.K., personal observations), we hypothesized that an environmental factor must have been moderating the underlying genetic susceptibility. As a test of this factor being related to diet, BTBR^{Ob} mice were maintained on the standard-chow diet fed in the breeding core (Teklad 2020X) instead of being switched to the standard-chow diet of our investigator-accessible facility (Purina 5001). These diets have similar energy densities and macronutrient contents but differ in some micronutrients and ingredient sources (Supplemental Table S1). Severe hyperglycemia rapidly developed, with 3 of 6 BTBR^{Ob} mice were euthanized due to endpoint criteria being reached before 10 weeks of age (Figure 1B).

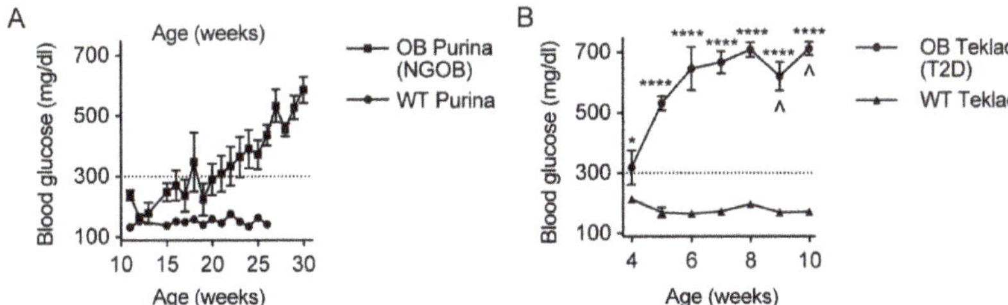

Figure 1. Diet partially explains the relative protection of a cohort of BTBR^{Ob} mice from early and severe Type 2 diabetes (T2D): (**A**) random-fed blood glucose readings of n = 6–8 Purina diet-fed wild-type (WT) and BTBR^{Ob} mice. Data represent mean ± SEM, although, as not all BG levels for all mice were recorded every week, a statistical analysis was not performed. (**B**) Weekly random-fed blood glucose readings of Teklad-fed WT and BTBR^{Ob} mice: n = 6–8 mice per group. Data represent mean ± SEM and were compared by two-way ANOVA with Holm-Sidak test post hoc. * $p < 0.05$ and **** $p < 0.0001$. ^ indicates that 1–2 BTBR^{Ob} mice were euthanized after human endpoints were reached.

2.2. An Altered Gut Microbiota Composition Is Associated with T2D Resistance vs. Susceptiblity BTBR^{Ob} Mice

Diet can influence gut microbiome composition, and recent discoveries have consistently demonstrated the role of the gut microbiome in T2D pathology [8]. In order to determine whether an intrinsic difference in the gut microbiome composition was related to T2D phenotype penetrance, we collected fecal pellets from three 11–17-week-old normoglycemic BTBR^{Ob} mice (NGOB) and three 4-week-old BTBR^{Ob} mice transferred directly from the breeding core facility (Pre-T2D). None of the mice were severely T2D: all of the NGOB mice and one Pre-T2D mouse had blood glucose levels <300 mg/dL, while the other two pre-T2D mice had blood glucose levels <400 mg/dL (Figure 2A). 16s rRNA sequencing was performed (a complete list of the relative abundance of microbiota by phyla, class, order, family and genus between NGOB and Pre-T2D mouse fecal pellets can be found in Supplemental Table S2). Briefly, Bacteroidetes and Firmicutes composed the majority of gut microbiota, as expected, with no significant differences between the groups (Figure 2B,C). While not statistically significant, the mean total nondominant phyla abundance was reduced in Pre-T2D vs. NGOB fecal samples (Figure 2D), an effect that appeared most related to Proteobacteria abundance (Figure 2E,F). Within Proteobacteria, the Gammaproteobacteria genus comprised half in the NGOB fecal samples and was absent in Pre-T2D samples. Finally, a three-way principal component analysis (PCA) analysis of the full dataset showed clustering by groups that, while not statistically significant, was certainly supportive of a complete study (Supplemental Table S2).

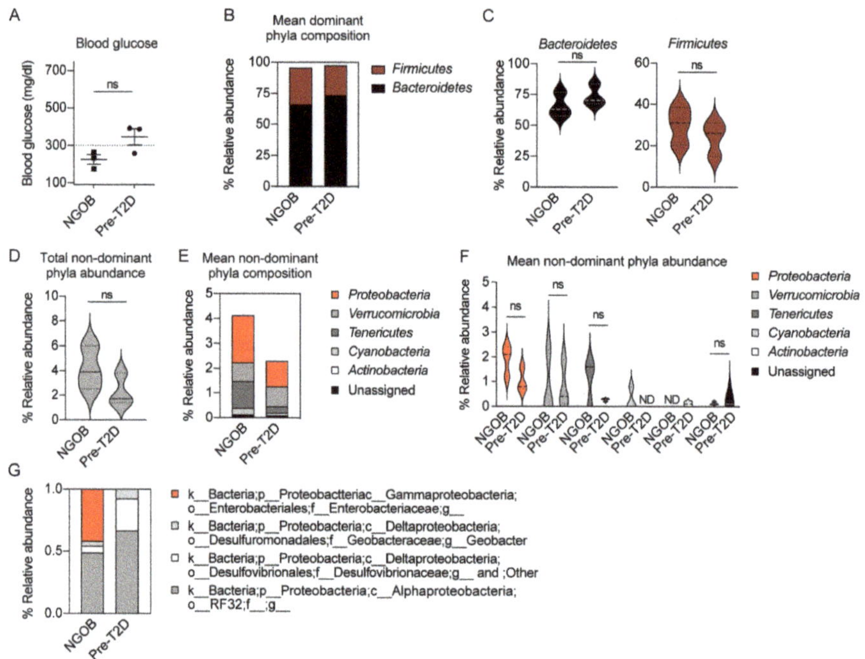

Figure 2. An altered gut microbiota composition is associated with T2D resistance vs. susceptibility BTBR[Ob] mice: (**A**) random-fed blood glucose levels of two female and one male mouse per group, where data represent mean ± SEM and ns represents not significant; (**B**) bar plot of dominant phyla; (**C**) violin plots of individual dominant phyla; (**D**) violin plots of total nondominant phyla; (**E**) bar plot of nondominant phyla; (**F**) violin plots of individual nondominant phyla; and (**G**) bar plot of individual proteobacteria orders as a fraction of the total proteobacteria abundance. Data represent mean ± SD and population density. In (**C,D,F**), data were compared by *t*-test. ns = not significant.

2.3. The Gut Microbiota Composition of Normoglycemic BTBROb Mice Is Significantly Different Compared to Both T2D BTBR[Ob] and WT Controls, At Least Partially Independent of Diet

In order to elucidate factors involved in T2D protection in the BTBR[Ob] line, new cohorts of mice were fed the Purina or Teklad diet from 4–10 weeks of age, at which time metabolic phenotyping and terminal tissue, blood, and cecal content collection were performed. As our pilot experiments included both male and female mice, these experiments were limited to the male sex to ensure robust hyperglycemia of Teklad-fed mice at 10 weeks of age. During the course of our full dietary intervention experiment, though, the T2D-protected phenotype of Purina-fed BTBR[Ob] mice disappeared. This provided us with the opportunity to explore factors involved in diabetes protection vs. susceptibility, at least partially independent of diet, as the WT and T2D groups were composed of both Purina-fed and Teklad-fed mice.

The mean 4–6 h fasting blood glucose levels of WT and NGOB mice were nearly identical (approximately 180 mg/dL), whereas that of T2D mice was approximately 500 mg/dL (Figure 3A). This effect was not due to differences in the Leptin[Ob] phenotype, as both

NGOB and T2D mice were similarly morbidly obese at 10 weeks of age compared to WT controls (Figure 3B). Furthermore, NGOB mice were equally as insulin-resistant as T2D littermates, with insulin having no reducing effect on blood glucose levels, whether represented as the percent change from baseline or as the area under the curve (AUC) from zero (Figure 3C).

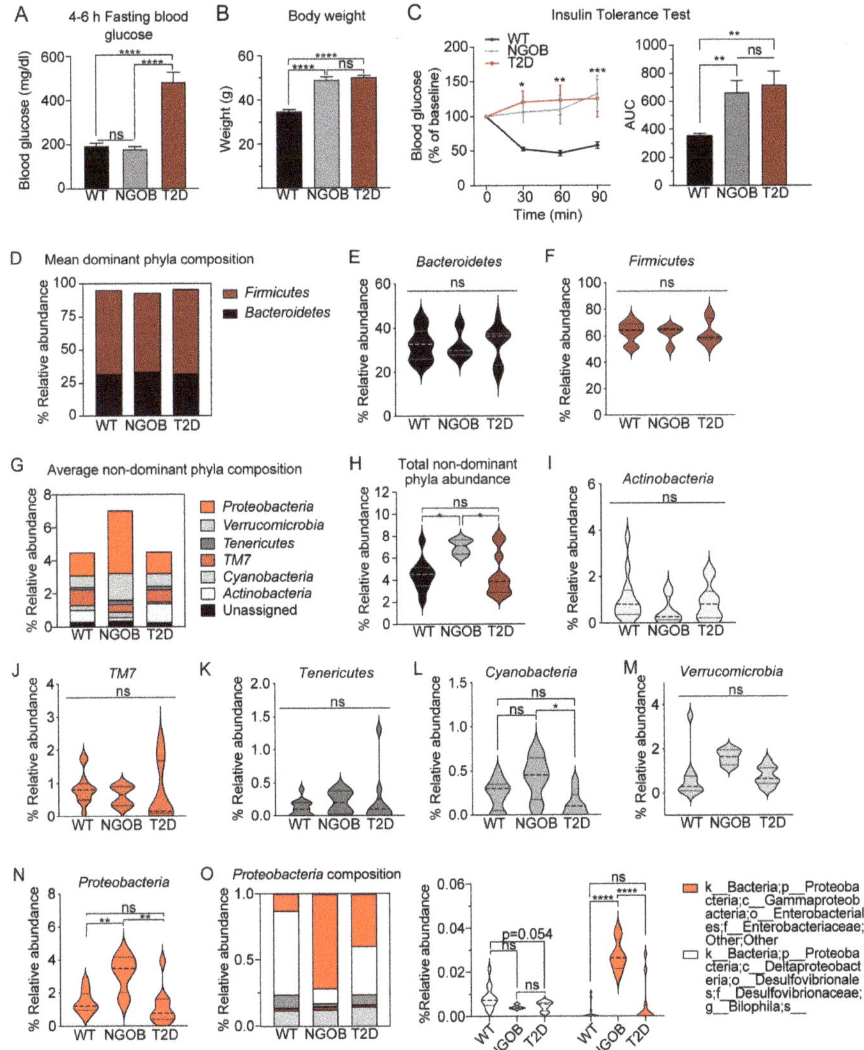

Figure 3. Gut microbiota diversity is associated with diabetes protection: (**A**) 4–6 h fasting blood glucose recordings of wild-type (WT) normoglycemic obese (NGOB) and diabetic (T2D) mice *n* = 6–15; (**B**) WT, NGOB, and T2D bodyweight at 10 weeks of age, *n* = 7–14; (**C**) insulin tolerance test of WT, NGOB, and T2D mice, with data plotted as a percent of baseline (t = 0 min) and quantified by the area under the curve (AUC) from zero and with data representing mean ± SEM, ns = not significant. (**D**) bar plot of dominant phyla; (**E**,**F**) violin plots of individual phyla; (**G**) bar plot of nondominant phyla; (**H**) average relative abundance of nondominant phyla; (**I–N**) violin plots of individual phyla; and (**O**) species composition of proteobacteria (left) and individual relative abundance of each proteobacteria species (right), *n* = 4–10. Data represent mean ± SD and population density. * *p* < 0.05, ** *p* < 0.01, *** *p* < 0.001, and **** *p* < 0.0001. ns = not significant.

16s rRNA sequencing was performed on cryopreserved cecal contents of 4–9 mice per group. The complete list of changes in microbial species (relative abundance) between WT, NGOB, and T2D mice and its statistical significance can be found in Supplementary Table S3. As with the pilot experiment using fecal pellets, Bacteroidetes and Firmicutes composed the majority of gut microbiota, with no significant differences among the groups (Figure 3D–F). Interestingly, phylogenetic diversity was similar between the WT and T2D groups but was significantly enhanced in NGOB animals (Figure 3G,H). Of the nondominant phyla, the relative abundance of Actinobacteria (Figure 3I), TM7 (Figure 3J), Tenericutres (Figure 3K), and Verrucomicrobia (Figure 3L) were all similar among WT, NGOB, and T2D mice. Cyanobacteria (Figure 3M) abundance was significantly lower in T2D mice compared to NGOB, and Proteobacteria abundance was substantially elevated in NGOB mice compared to both of the other groups (Figure 3N). In addition to altered total abundance, the Proteobacteria species composition in NGOB mice was vastly different compared to that of WT and T2D mice (Figure 3O, left), with specific changes in the Desulfovibrionaceae and Enterobacteriaceae families (Figure 3O, right).

2.4. The T2D Phenotype Is Associated with Altered Circulating Incretin and Adipokine Levels

Peptide hormones secreted from a number of tissues are critical for appropriate blood glucose control, and their levels are known to be dysregulated in IR and T2D [9–13]. Consistent with insulin hypersecretion to compensate for peripheral IR, random-fed plasma insulin levels were significantly elevated in NGOB and T2D mice compared to WT controls, with a lower mean plasma insulin level in T2D mice suggestive of emerging β-cell failure (Figure 4A).

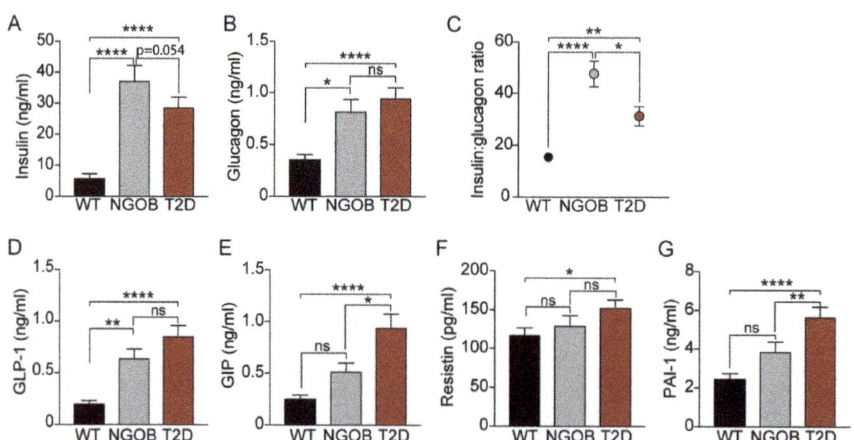

Figure 4. Loss of the insulin-glucagon ratio is associated with elevated incretin and adipokine levels in T2D: random-fed plasma concentrations of (**A**) insulin and (**B**) glucagon; (**C**) insulin-glucagon ratio; and random-fed plasma concentrations of (**D**) glucagon-like peptide 1 (GLP-1), (**E**) gastric inhibitory polypeptide (GIP), (**F**) resistin, and (**G**) gastric inhibitory polypeptide (PAI-1) in WT, NGOB, and T2D mice, n = 9–17. Data represents mean ± SEM. * $p < 0.05$, ** $p < 0.01$, and **** $p < 0.0001$. ns = not significant.

Failure to suppress α-cell glucagon secretion is another hallmark of IR, and consistent with this, NGOB and T2D plasma glucagon levels were similarly elevated compared to WT controls (Figure 4B). A high insulin-glucagon ratio (IGR) promotes glycogenolysis and gluconeogenesis while the opposite is true for low IGR [14]. Consistent with this concept, when each mouse's insulin value was normalized to its own glucagon value, IGR was significantly elevated in the NGOB group compared to both the WT and T2D littermates (Figure 4C). Gut incretin hormones such as glucagon-like peptide 1 (GLP-1) and gastric inhibitory polypeptide (GIP) are classically defined as potentiators of GSIS, and defects in their signaling pathways are known to contribute to the inappropriate glucose control of T2D [15]. Consistent with this concept, both GLP-1 and GIP were elevated in NGOB plasma compared to WT and even further elevated in the T2D state (Figure 4D,E). Finally, alterations in adipokine secretion due to adipose tissue meta-inflammation is becoming increasingly understood as contributing to the pathophysiology of IR and T2D [16]. The adipokines resistin and plasminogen activator inhibitor-1 (PAI-1) were both significantly elevated in T2D plasma compared to WT (Figure 4F,G). PAI-1 was mildly elevated in NGOB plasma, albeit not with statistical significance (Figure 4G).

2.5. An Integrated Ultrahigh-Resolution FIE-FTCIR MS Metabolomics Approach Accurately Clusters Mouse Plasma Samples by Disease State

We performed untargeted plasma metabolomics using a workflow integrating flow injection electrospray (FIE) with ultrahigh-resolution Fourier Transform Ion Cyclotron Resonance (FTICR) mass spectrometry (MS), a platform recently established and validated for use with cryopreserved plasma samples [7] (see Supplementary Figure S1 for a summary of the workflow). Between 1200 and 2200 distinct metabolic features were detected in plasma samples from each group, with over 75% of those features being annotatable by chemical structure using MetaboScape or chemical name using the METLIN Metabolite and Chemical Entity Database (Figure 5A and Supplementary Table S4). Of the 1750 distinct metabolic features, 653 were shared among the three groups (Figure 5B). A principal component analysis (PCA) performed using all detected metabolic features in both positive and negative ion modes was used to assess variability in the data (Figure 5C). Unsupervised hierarchical clustering effectively grouped plasma samples by phenotype (Figure 5D, top), with a weaker effect of diet (Figure 5D, bottom). Finally, using METLIN-annotated data, a robust increase in intensity for a hexose corresponding with a mass of 180.06351 was found in T2D plasma compared to NGOB and WT (Figure 5E): a peak we confirmed by MS/MS in a previous work to be primarily glucose, with a possible minor contribution by fructose [7].

2.6. Diet Alone Does Not Explain the Phenotype of Male BTBR^{ob} Mice at 10 Weeks of Age

Pairwise comparisons of the metabolic features of WT vs. NGOB plasma and NGOB vs. T2D plasma were performed to further validate the phenotypic predictive value of our analysis. Volcano plots were generated to elucidate significantly differentially expressed features (Figure 6A,B, left), and the top 25 most highly differentially expressed of these were again used in unsupervised hierarchical clustering analyses (full lists of the differentially expressed features between WT and NGOB plasma vs. NGOB vs. T2D plasma are shown in Supplemental Tables S5 and S6, respectively). In these pairwise comparisons, an even more robust clustering of plasma samples by phenotypic group was found, with complete segregation of WT from NGOB samples and of NGOB from T2D samples (Figure 6A,B, right, top). Furthermore, while all obese mice that remained normoglycemic had been fed the Purina diet, diet alone was not predictive of phenotype, as nearly equal numbers of T2D mice had been fed the Purina vs. Teklad diet, and even in pairwise comparisons, T2D mouse plasma samples did not segregate exclusively by diet group (Figure 6B, right, bottom).

2.7. FIE-FTCIR MS Metabolomics Reveals That Elevations in Circulating Eicosanoid Precursors Correlate Directly with Agonist-Dependent EP3 Signaling in T2D β-Cell Dysfunction

In our previously published work [7], the fatty acids and fatty acid conjugates family was the most highly significantly expressed for both polar and nonpolar metabolites in plasma from obese mice [7]. Eicosanoids are highly bioactive fatty acid metabolites, and a subset of essential polyunsaturated fatty acids (PUFAs) provides the substrates for their synthesis [17,18]. Using the METLIN-annotated metabolic features shown in Supplemental Tables S3 and S4, the entire linoleic acid elongation and desaturation pathway was downregulated in NGOB vs. WT plasma: in most cases, with statistical significance (Figure 7A, left, gray vs. black bars). Furthermore, in all cases, the abundance of these omega-6 PUFAs was significantly elevated compared to both the other groups (Figure 7A, left, red bars). Linolenic acid serves as the initial backbone for the omega-3 PUFA subfamily, and while linolenic acid cannot be distinguished from the omega-6 PUFA gamma linolenic acid (GLA) in our analysis, its downstream products, eicosapentaenoic acid (EPA) and clupanodonic acid/docosapentaenoic acid, were also significantly upregulated in T2D plasma vs. WT (Figure 7A, right, red vs. black bars). In these cases, though, their levels in NGOB plasma were not lower than in WT, and in fact, EPA abundance was significantly elevated (Figure 7A, right).

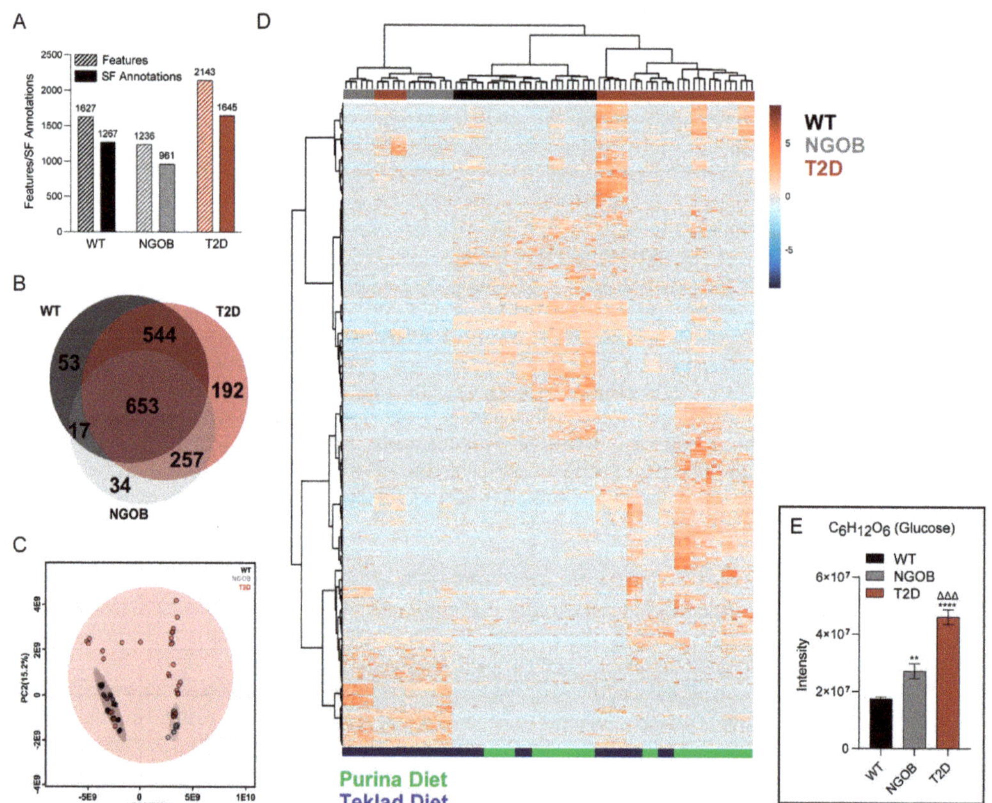

Figure 5. Application of flow injection electrospray (FIE)-Fourier Transform Ion Cyclotron Resonance (FTICR) mass spectrometry (MS) to plasma samples from WT, NGOB, and T2D mice: (**A**) the numbers of features (light shading) and chemical formula annotations (solid shading) from MetaboScape 4.0 ($\Delta m < 5$ ppm) of the WT, NGOB, and T2D groups; (**B**) Venn diagram of chemical formula annotations of WT, NGOB, and T2D plasma samples; (**C**) Principal component analysis (PCA) of the three groups based on FIE-FTICR MS data, where the 95% confidence limit is indicated as the shaded area; (**D**) heat map of significantly expressed metabolic features annotated by a chemical formula generated by unsupervised hierarchical clustering analysis; and (**E**) intensity of the hexose peak, experimentally confirmed to be primarily glucose, in WT, NGOB, and T2D plasma. Data represent mean ± SEM and were compared by one-way ANOVA with Holm-Sidak test post hoc. ** $p < 0.01$ and **** $p < 0.0001$ compared to WT. ΔΔΔ $p < 0.001$ compared to NGOB.

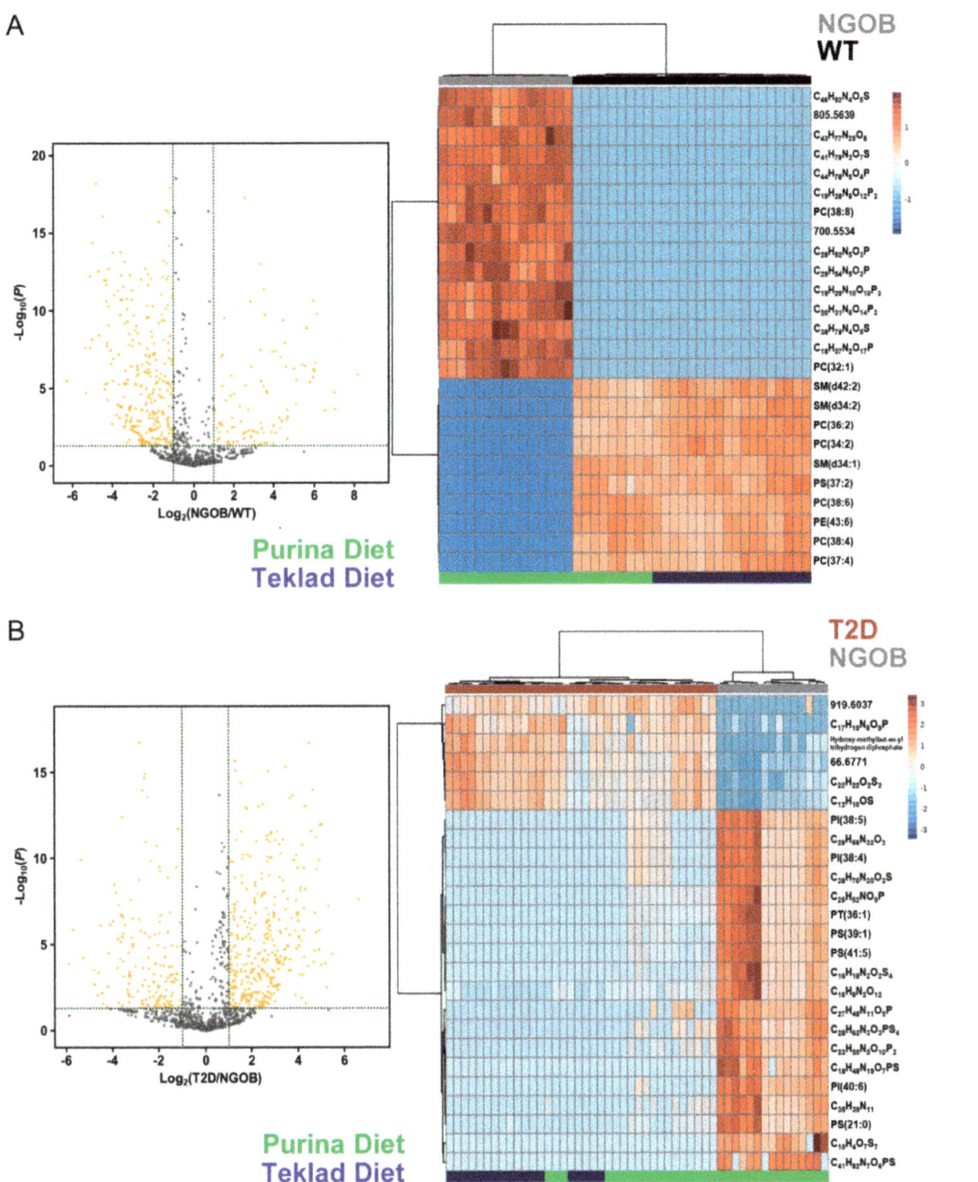

Figure 6. Volcano plots used to determine significantly differentially expressed annotated metabolic features (left) and heat maps created by unsupervised hierarchical clustering using the top 25 most highly differentially expressed of these (right) in pairwise comparisons of (**A**) NGOB vs. WT plasma samples and (**B**) T2D vs. NGOB plasma samples.

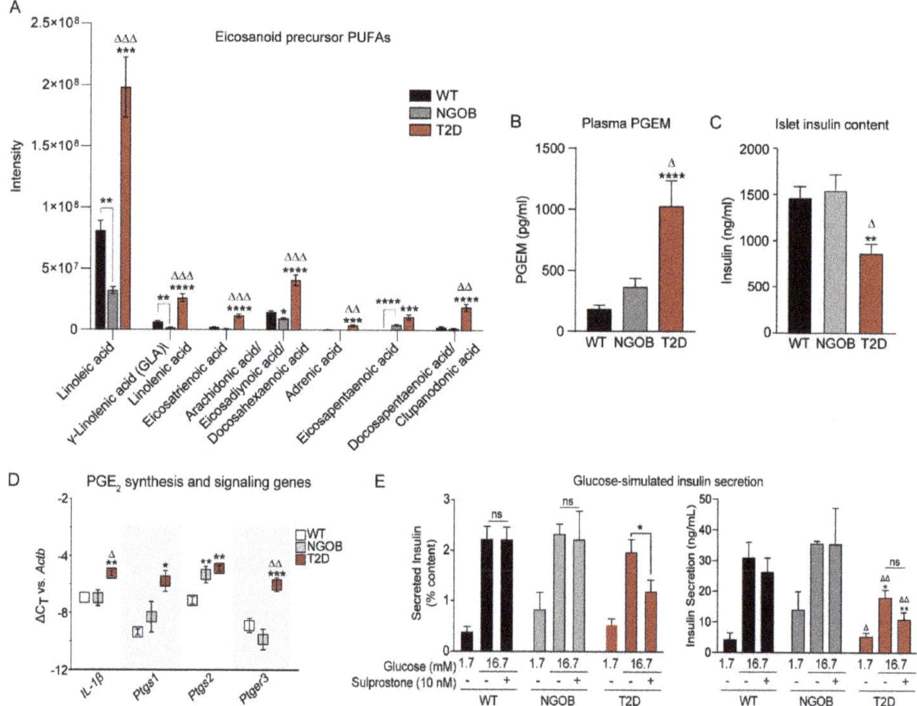

Figure 7. Changes in PGE₂ production and signaling and its role in insulin secretion from WT, NGOB, and T2D BTBR mice: (**A**) comparison of significantly expressed eicosanoid precursors from WT, NGOB, and T2D plasma samples, where data were analyze by multiple *t*-test with Holm-Sidak test post hoc; (**B**) plasma PGE metabolite concentrations as measured by ELISA (n = 4–12); (**C**) total islet insulin content from WT, NGOB, and T2D islets, n = 3–12; (**D**) transcript expression of PGE₂ synthetic and signaling enzymes from WT, NGOB, and T2D islets, where data are relative to β-actin, n = 4–6; and (**E**) glucose-stimulated insulin secretion from islets stimulated with 1.7 mM glucose or 16.7 mM glucose ±10 nM sulprostone. Secreted insulin was normalized to total insulin content (left). Data are plotted as total insulin secreted (right), n = 3–12. Data represent mean ± SEM and were analyzed by two-way ANOVA with Holm-Sidak test post hoc. * $p < 0.05$, ** $p < 0.01$, *** $p < 0.001$, and **** $p < 0.0001$ compared to WT. Δ $p < 0.05$, ΔΔ $p < 0.001$, and ΔΔΔ $p < 0.001$. ns = not significant.

The EP3 agonist prostaglandin E₂ (PGE₂), a metabolite of arachidonic acid (AA) incorporated into plasma membrane phospholipids, and PGE₂ excretion have been found to be upregulated in islets from T2D mice and humans compared to nondiabetic controls [3,19]. EPA competes with AA for the same phospholipid position and downstream metabolic enzymes. While PGE₂ is rapidly degraded in the blood, a specific prostaglandin E metabolite (PGEM) ELISA confirmed that elevated conjugated fatty acid levels correlated directly with increased circulation of PGE, of which PGE₂ is by far the most abundant (Figure 7B). Finally, the baseline insulin content of islets from T2D mice was significantly reduced compared to both of the other groups (Figure 7C), consistent with β-cell failure downstream of a failed β-cell compensation response [4,6].

In the β-cell, the pro-inflammatory cytokine interleukin-1β (IL-1β) is known to stimulate the expression of both PGE₂ synthetic enzymes and EP3 itself [20–23]. While IL-1β mRNA expression was nearly identical in islets from NGOB mice compared to WT, it was significantly upregulated in islets from T2D mice (Figure 7D, left). Similarly, the expression of mRNAs encoding cyclooxygenase 1 and 2 (Ptgs1 and Ptgs2, respectively), catalyzing the rate limiting step in PGE₂ synthesis, was also significantly upregulated in islets from T2D mice compared to WT (with the latter also being higher in NGOB) (Figure 7D, middle).

Finally, EP3 (Ptger3) mRNA expression was only upregulated in T2D compared to both WT and NGOB, with a fold-change of approximately 16-fold when calculated via the $2^{\Delta\Delta Ct}$ method (Figure 7D, right).

To confirm altered PGE_2/EP3 signaling as a direct contributor to β-cell dysfunction in this T2D mouse model, we performed ex vivo glucose-stimulated insulin secretion (GSIS) assays with and without the addition of an EP3-selective agonist, sulprostone. Normalizing the amount of insulin secreted to total insulin content revealed enhanced insulin secretion at low glucose (1.7 mM) in NGOB islets compared to WT, consistent with the hypersecretory phenotype required to maintain fasting euglycemia, with no effect of sulprostone (Figure 7E, gray vs. black bars). In contrast, islets from T2D mice have a mild GSIS defect in response to stimulatory (16.7 mM) glucose that is even further exaggerated with sulprostone treatment (Figure 7E, red bars).

3. Discussion

To date, the BTBR^ob mouse has been understood as a model of severe T2D secondary to β-cell failure, with a consistent, full disease penetrance by 16 weeks of age in both sexes and earlier in males [4,6]. In this study, we encountered a unique cohort of male BTBR^ob mice for which environmental factors, including but not limited to diet, completely prevented hyperglycemia without further intervention. We have exploited these findings to study physiological changes as they pertain to islet function in obese, insulin-resistant, normoglycemic animals compared to their equally obese and insulin-resistant T2D littermates.

The gut microbiome is of great interest in the T2D field and is well-known to be related to diet and other environmental conditions. Previous reports on mice and humans have documented alterations in the gut microbiome of T2D individuals compared to healthy controls [8,24]. While changes in the ratio of the dominant phyla, *Firmicutes* and *Bacteroidetes*, have been previously linked with T2D, we found no differences in *Firmicutes* or *Bacteroidetes* composition or relative abundance among groups [24,25]. Instead, the most defining characteristic of the NGOB gut microbiome fingerprint compared to both of the other groups was a change in nondominant phyla composition and abundance driven primarily by *Proteobacteria*. The elevations in *Proteobacteria* in NGOB mouse cecal matter are perplexing, as many reports indicate a higher disease incidence when *Proteobacteria* is elevated [26]. On the other hand, a richer overall microbial diversity has been associated with positive outcomes regarding glucose homeostasis [8,27].

A distinct endocrine profile was observed in the NGOB mouse compared to WT, and this profile was further altered in T2D. Elevated insulin and glucagon are hallmarks of β-cell stress and insulin resistance, and NGOB mice showed clear β-cell compensation, with a robust increase in IGR that was lost in the T2D cohort. The gut-islet relationship, of great interest in the T2D field, is thought to be modulated in part by the microbial composition of the GI tract. The incretin hormones, GLP-1 and GIP, are primarily secreted from gut enteroendocrine cells and act on the β-cell to promote insulin secretion. GLP-1 and GIP were both elevated in NGOB mice compared to WT. Surprisingly, they were even further elevated in plasma from T2D mice, even though loss of the incretin response is a known pathophysiological defect in T2D [27]. However, gut microbiota alterations have been shown to influence incretin sensitivity in obesity, insulin resistance, and T2D [28,29]; therefore, elevated GLP-1 and GIP levels may be reflective of a compensatory response that ultimately fails to promote β-cell function.

Changes in gut microbiota are also associated with alterations in circulating factors associated with T2D pathology [28,30,31]. This relationship with microbiota dysbiosis is further supported by elevated levels of resistin and PAI-1, adipokines associated with obesity and T2D [30,32,33]. Resistin and PAI-1 production and secretion are stimulated during innate immune response and, therefore, can reflect alterations in gut endothelial cell integrity that impact the systemic metabolic profile of the organism [30,33–36]. In obesity, bacterial byproducts, such as lipo-polysaccharides (LPS), enter circulation as the endothelial wall of the intestine is degraded: a common consequence of obesity-associated intestinal inflammation [37,38]. LPS induces the release of resistin and PAI-1 from adipocytes, and LPS has also been shown to impair β-cell function via downregulation of the mature β-cell gene expression profile [39,40]. Therefore, it is possible that resistin and PAI-1 may serve as markers for β-cell dysfunction of T2D, independent of any direct influence on the β-cell themselves. More work would be necessary to tease apart this relationship.

The effects of elevated plasma AA and its precursor linoleic acid have been previously implicated in diabetes pathology and correlate with elevated HBA1c levels and hyperglycemia in human subjects [41]. In this work, we confirmed that elevated levels in circulating AA (or its isomers) correlate directly with elevated PGE metabolite levels. However, it has long been known that PGE_2 is synthesized and excreted from pancreatic β-cells themselves and that this synthesis is a factor of both substrate availability and the expression of key synthetic enzymes, many of which are upregulated by the pro-inflammatory cytokine IL-1β [3,4,20–22]. In this work, we confirmed that this direct IL-1β-associated effect is of biological consequence, as the PGE_2 analog, sulprostone, only influences the GSIS response of islets from T2D BTBR^{Ob} mice, which already exhibits a prominent secretion defect compared to their nondiabetic controls. However, even though essential PUFAs must be obtained from the diet, diet alone cannot explain the altered ratios of omega-6 and omega-3 PUFAs downstream of linoleic acid and linolenic acid (including the most important to our work, AA and EPA) as equal numbers of T2D mice had also been fed an identical diet. Still, when considered in the context of our previous work demonstrating that BTBR^{Ob} islets cultured in EPA-enriched media or nonobese diabetic mice fed an EPA-enriched diet significantly improved ex vivo or in vivo β-cell function, respectively [3], our findings certainly support the continued study of PUFA-based dietary interventions for T2D prevention or therapy.

In summary, we found that a host of systemic metabolic alterations are associated with the T2D phenotype of a strong genetic model of the disease and that, at least for the AA-to-PGE_2 pathway, have directly implicated a specific class of metabolites in β-cell dysfunction of the disease. These findings have strong implications in the management of T2D, as many foods in the Western diet are enriched with AA. Therefore, pharmacological strategies to reduce gut/adipose inflammation in combination with a diet focused on limiting circulating AA may help to facilitate better β-cell function, either alone or in concert with T2D medications, promoting more effective blood glucose control even in the face of chronic obesity and insulin resistance.

4. Materials and Methods

4.1. Animal Care and Husbandry

BTBR mice heterozygous for the *Leptin^{ob}* mutation were purchased from The Jackson Laboratory and bred in house at the UW-Madison Breeding Core Facility to generate homozygous OB mice or wild-type (WT) controls. Experimental mice were singly housed in temperature- and humidity-controlled environments and maintained on a 12:12 h day/light cycle with free access to acidified water (InnoVive, San Diego, CA, USA) and one of two standard mouse chows of nearly identical macronutrient composition and energy density. A direct comparison of the nutrient content and ingredients of the Teklad global soy protein-free extruded 2920X diet (Envigo, Indianapolis, IN, USA) or the Rodent Laboratory Chow 5001 diet (Purina, Neenah, WI, USA) has been previously published [7]. Both wild-type control mice ($n = 9$; 3-Purina and 6-teklad) and OB mice ($n = 17$; 11-Purina and 6-teklad) were given either Purina or Teklad diets. At 10 weeks of age, OB mice that developed hyperglycemia were grouped as T2D/HGOB ($n = 12$). A subset of OB mice was able to maintain normoglycemia, which was grouped as NGOB ($n = 5$). All protocols were approved by the Institutional Animal Care and Use Committees of the University of Wisconsin-Madison and by the William S. Middleton Memorial Veterans Hospital, which are both accredited by the Association for Assessment and Accreditation of Laboratory

Animal Care (Project ID: G005181-R01-A03). All animals were treated in accordance with the standards set forth by the National Institutes of Health Office of Animal Care and Use.

4.2. Blood Glucose Measurements and Insulin Tolerance Tests

Insulin tolerance tests (ITTs) were performed essentially as previously described [42]. Briefly, mice were fasted for 4–6 h and 0.75 U/kg short-acting recombinant human insulin (Humulin® R; Eli Lilly, Indianapolis, IN, USA) was injected intraperitoneally. Blood glucose readings were taken by tail nick with an AlphaTRACK glucometer (Zoetis, Parsippany-Troy Hills, NJ, USA) at baseline and the indicated times after insulin administration. Percent blood glucose change from baseline was determined by normalizing the blood glucose reading at each timepoint to that at baseline for each mouse independently. The blood glucose percent change from baseline was averaged within genotypes at each time point, giving the means $\pm$ SEM.

4.3. Ex Vivo Islet Glucose Stimulated Insulin Secretion Assays

Islets were isolated from experimental mice at 10 weeks of age utilizing a collagenase digestion protocol as previously described [43]. Glucose-stimulated insulin secretion (GSIS) assays were performed as previously described [44] after the indicated treatments. Briefly, islets were washed and preincubated for 45 min in a 0.5% Bovine Serum Albumin (BSA) Krebs Ringer Bicarbonate Buffer (KRBB) containing 1.7 mM glucose. Islets were then incubated for an additional 45 min in either low glucose (1.7mM) or stimulatory glucose (16.7 mM) $\pm$ the EP3-selective agonist sulprostone (10 nM). Secretion media was collected, and islets were lysed in equal volume to determine insulin content. Insulin was measured via ELISA as previously described [44].

4.4. Quantitative PCR for Gene Expression Analyses

RNA isolation, cDNA synthesis, and quantitative PCR using SYBR Green reagent (Bio-Rad) to determine relative mRNA abundance among groups were all performed as previously described [45]. Data were normalized to that of β-actin to calculate ΔCT values. Primer sequences are available upon request.

4.5. Terminal Blood Collection and Plasma Hormone/Metabolite Assays

Terminal blood collection for plasma samples was performed by retroorbital puncture under anesthesia. Briefly, mice were anesthetized using 2,2,2-tribromoethanol (Sigma-Aldrich, St. Louis, MO, USA; #T48402). Blood was collected retro-orbitally using a heparin-coated glass capillary tube and mixed with 5 µM EDTA, 10 nM DDP-4 inhibitor, and 20 nM aprotinin. Plasma was isolated via centrifugation and stored at $-80\,°C$ until needed. Plasma hormones were measured utilizing the Bio-Plex Pro™ Diabetes Assay (Bio-Rad Laboratories, Hercules, CA, USA) following the manufacturer's protocol. The plasma prostaglandin E metabolite was measured utilizing a PGEM ELISA (Cayman Chemical Company, Ann Arbor, MI, USA; no. 514531) following the manufacturer's protocol as previously described [42].

4.6. Gut Microbial DNA Preparation, Sequencing, and Analysis

Microbial 16s sequencing of cecal matter was performed as previously described [46]. Briefly, 20–50 mg of the cecal matter was collected from WT, NGOB, and T2D mice, and genomic DNA was extracted and cleaned by using the Macherey-Nagel PCR Clean-up kit according to manufacturer's protocol (ThermoFisher Scientific, Waltham, MA, USA). Purified genomic DNA was submitted to the University of Wisconsin-Madison Biotechnology Center. DNA concentration was verified fluorometrically, and samples were prepared and amplified according to Illumina's 16s Metagenomic Sequencing Library Preparation Protocol with few modifications as described before [46]. Following PCR, reactions were cleaned using $0.7\times$ volume of AxyPrep Mag PCR clean-up beads (Corning, Corning, NY, USA). Quality and quantity of the finished libraries were assessed using an Agilent DNA

1000 kit (Agilent Technologies, Santa Clara, CA, USA) and Qubit® dsDNA HS Assay Kit (ThermoFisher Scientific), respectively. In an equimolar fashion, libraries were pooled and appropriately diluted prior to sequencing. After sequencing, images were analyzed using the standard Illumina Pipeline, version 1.8.2. OTU assignments (Illumina, Inc., San Diego, CA, USA). Diversity plots were created using the QIIME (Ver. 1.9.1) analysis pipeline [46,47].

For the pilot experiment using fecal pellets from NGOB and pre-T2D mice, cryopreserved fecal pellets were shipped to Argonne National Labs where sequencing and data analysis were performed according to their standard protocols. Sequencing from the pilot experiment only went to the order level and not the species level.

4.7. FIE-FTCIR MS for Unbiased Plasma Metabolomics

Plasma samples were collected as described in Section 4.5. Samples were prepared using a methanol extraction procedure, and unbiased FIE-FTICR MS experiments were performed as previously described [7]. Statistical analysis was performed using MetaboScape 4.0 (Bruker, Billerica, MA, USA) and the online software MetaboAnalyst [48]. Putative metabolites were further annotated by METLIN with a 2-ppm mass error cutoff. Details about data analysis were described previously [7].

4.8. Statistical Analyses

Data from all experiments excluding that previously described for microbiome and metabolomics analyses were analyzed using GraphPad Prism v.9 (GraphPad Software Inc., San Diego, CA, USA). Data were analyzed by t-test, one-way ANOVA, or two-way ANOVA as described in the figure legends. $p < 0.05$ was considered statistically significant.

4.9. Data Availability

All data contained within this manuscript are available upon reasonable request of the corresponding author. The mass spectrometry proteomics data have been deposited to the ProteomeXchange Consortium via the PRIDE [49] partner repository with the dataset identifier PXD022624.

Supplementary Materials: The following are available online at https://www.mdpi.com/2218-1989/11/1/58/s1, Figure S1: Workflow of the FIE-FTICR MS-based platform for metabolomics. Table S1: Diet Composition, Table S2: Pilot 16s rRNA sequencing, Table S3: 16S rRNA analysis, Table S4: FIE-FTICR MS Mass List, Table S5: WT vs NGOB metabolites, Table S6: NGOB vs T2D metabolites.

Author Contributions: Conceptualization, M.D.S., Y.Z., J.C.N., B.T.L., D.W.L., Y.G. and M.E.K.; data curation, Y.Z., Y.G. and M.E.K.; formal analysis, M.D.S., Y.Z., N.E.R., C.P., I.M.O., R.J.F., J.C.N., E.G., A.R., H.K.S. and M.E.K.; funding acquisition, I.M.O., E.D.C., D.B.D., B.T.L., A.R.B., D.W.L., Y.G. and M.E.K.; investigation, M.D.S., Y.Z., N.E.R., I.M.O., R.J.F., J.C.N., E.G., A.R., H.K.S., M.H.F. and M.E.K.; methodology, M.D.S., Y.Z., N.E.R., D.W.L., Y.G. and M.E.K.; project administration, E.D.C., D.B.D. and M.E.K.; supervision, M.D.S., B.T.L., A.R.B., D.W.L., Y.G. and M.E.K.; visualization, M.D.S., Y.Z., N.E.R. and C.P.; writing—original draft, M.D.S., N.E.R., C.P. and J.C.N.; writing—review and editing, M.D.S., Y.Z., C.P., R.J.F., A.R., D.B.D., B.T.L., D.W.L., Y.G. and M.E.K. All authors have read and agreed to the published version of the manuscript.

Funding: This work was supported in part by Merit Review Awards I01 BX003700 (to M.E.K.), I01 BX004031 (to D.W.L.), and I01 BX003382 (to B.T.L.) from the United States (U.S.) Department of Veterans Affairs Biomedical Laboratory Research and Development (BLR&D) Service. This work was also supported in part by National Institutes of Health (NIH) grants K01 DK080845 (to M.E.K.), R01 DK102598 (to M.E.K.), R01 AG056771 (to D.W.L.), UL1 TR002373 (to A.R.B.), R01 DK104927 (to B.T.L.), R01 DK111848 (to B.T.L.), U01 DK127378 (to B.T.L.), P30 DK020595 (to B.T.L.), and P30 CA014520 (to I.M.O.); American Diabetes Association Grant 1-14-BS-115 (to M.E.K.); and a UW2020 grant from the Wisconsin Alumni Research Foundation (to M.E.K., E.D.C., and D.B.D). This research was conducted while D.W.L. was an AFAR Research Grant recipient from the American Federation for Aging Research. N.E.R. and J.C.N. were supported in part by training grants from the UW Institute on Aging (NIA T32 AG000213).

Institutional Review Board Statement: The study was conducted according to the guidelines of the Declaration of Helsinki, and approved by the Institutional Animal Care and Use Committees of the University of Wiscon-sin-Madison and by the William S. Middleton Memorial Veterans Hospital (protocol code: G005181-R01-A03 and date of approval: 6/13/2018).

Informed Consent Statement: Not applicable.

Data Availability Statement: All data are contained within the article or supplementary material.

Conflicts of Interest: M.D.S., Y.Z., N.E.R., C.P., I.M.O., R.J.F., E.G., H.K.S., M.H.F., E.D.C., D.B.D., B.T.L., A.R.B., Y.G. and M.E.K. declare that they have no conflicts of interest with the contents of this article. J.C.N. is now employed by Novo Nordisk (800 Scudders Mill Rd, Plainsboro, NJ 08536). This work was completed in full during his predoctoral training with M.E.K. and is not related to his current position. D.W.L. received funding from and is a scientific advisory board member of Aeovian Pharmaceuticals, which seeks to develop novel, selective mTOR inhibitors for the treatment of various diseases, including diabetes. The content is solely the responsibility of the authors and does not necessarily represent the official views of the National Institutes of Health, the U.S. Department of Veterans Affairs, or the United States Government. The funders had no role in the design of the study; in the collection, analyses, or interpretation of data; in the writing of the manuscript; or in the decision to publish the results.

References

1. Cerf, M.E. Beta cell dysfunction and insulin resistance. *Front. Endocrinol.* **2013**, *4*. [CrossRef] [PubMed]
2. Kasuga, M. Insulin resistance and pancreatic β cell failure. *J. Clin. Investig.* **2006**, *116*, 1756–1760. [CrossRef] [PubMed]
3. Neuman, J.C.; Schaid, M.D.; Brill, A.L.; Fenske, R.J.; Kibbe, C.R.; Fontaine, D.A.; Sdao, S.M.; Brar, H.K.; Connors, K.M.; Wienkes, H.N.; et al. Enriching islet phospholipids with eicosapentaenoic acid reduces prostaglandin E_2 signaling and enhances diabetic β-cell function. *Diabetes* **2017**, *66*, 1572–1585. [CrossRef] [PubMed]
4. Kimple, M.E.; Keller, M.P.; Rabaglia, M.R.; Pasker, R.L.; Neuman, J.C.; Truchan, N.A.; Brar, H.K.; Attie, A.D. Prostaglandin E_2 receptor, EP3, is induced in diabetic islets and negatively regulates glucose- and hormone- stimulated insulin secretion. *Diabetes* **2013**, *62*, 1904–1912. [CrossRef] [PubMed]
5. Bi, C.; Fu, Y.; Zhang, Z.; Li, B. Prostaglandin E_2 confers protection against diabetic coronary atherosclerosis by stimulating M2 macrophage polarization via the activation of the CREB/BDNF/TrkB signaling pathway. *FASEB J.* **2020**, *34*, 7360–7371. [CrossRef] [PubMed]
6. Clee, S.M.; Nadler, S.T.; Attie, A.D. Genetic and genomic studies of the BTBR ob/ob mouse model of type 2 diabetes. *Am. J. Ther.* **2005**, *12*, 491–498. [CrossRef] [PubMed]
7. Zhu, Y.; Wancewicz, B.; Schaid, M.; Tiambeng, T.N.; Wenger, K.; Jin, Y.; Heyman, H.; Thompson, C.J.; Barsch, A.; Cox, E.D.; et al. Ultrahigh-Resolution Mass Spectrometry-Based Platform for Plasma Metabolomics Applied to Type 2 Diabetes Research. *J. Proteome Res.* **2020**, *20*, 463–473. [CrossRef] [PubMed]
8. Harsch, I.; Konturek, P. The Role of Gut Microbiota in Obesity and Type 2 and Type 1 Diabetes Mellitus: New Insights into "Old" Diseases. *Med. Sci.* **2018**, *6*, 32. [CrossRef]
9. Falinska, A.M.; Tan, T.; Bloom, S. Gut hormones and Type 2 diabetes mellitus. *Diabetes Manag* **2014**, 501–513. [CrossRef]
10. Lu, H.L.; Wang, H.W.; Wen, Y.; Zhang, M.X.; Lin, H.H. Roles of adipocyte derived hormone adiponectin and resistin in insulin resistance of type 2 diabetes. *World J. Gastroenterol.* **2006**, *12*, 1747–1751. [CrossRef]
11. Rorsman, P.; Huising, M.O. The somatostatin-secreting pancreatic δ-cell in health and disease. *Nat. Rev. Endocrinol.* **2018**, *14*, 404–414. [CrossRef] [PubMed]
12. Hædersdal, S.; Lund, A.; Knop, F.K.; Vilsbøll, T. The Role of Glucagon in the Pathophysiology and Treatment of Type 2 Diabetes. *Mayo Clin. Proc.* **2018**, *93*, 217–239. [CrossRef] [PubMed]
13. Fu, Z.; Gilbert, E.R.; Liu, D. Regulation of Insulin Synthesis and Secretion and Pancreatic Beta-Cell Dysfunction in Diabetes. *Curr. Diabetes Rev.* **2012**, *9*, 25–53. [CrossRef]
14. Kalra, S.; Gupta, Y. The Insulin:Glucagon Ratio and the Choice of Glucose-Lowering Drugs. *Diabetes Ther.* **2016**, *7*. [CrossRef] [PubMed]
15. Drucker, D.J. The biology of incretin hormones. *Cell Metab.* **2006**, *3*, 153–165. [CrossRef]
16. Kwon, H.; Pessin, J.E. Adipokines mediate inflammation and insulin resistance. *Front. Endocrinol.* **2013**, *4*. [CrossRef]
17. Khanapure, S.; Garvey, D.; Janero, D.; Letts, G. Eicosanoids in Inflammation: Biosynthesis, Pharmacology, and Therapeutic Frontiers. *Curr. Top. Med. Chem.* **2007**, *7*, 311–340. [CrossRef]
18. Zhou, L.; Nilsson, Å. Sources of eicosanoid precursor fatty acid pools in tissues. *J. Lipid Res.* **2001**, *42*, 1521–1542.
19. Park, J.Y.; Pillinger, M.H.; Abramson, S.B. Prostaglandin E_2 synthesis and secretion: The role of PGE_2 synthases. *Clin. Immunol.* **2006**, *119*, 229–240. [CrossRef]

20. Parazzoli, S.; Harmon, J.S.; Vallerie, S.N.; Zhang, T.; Zhou, H.; Robertson, R.P. Cyclooxygenase-2, not microsomal prostaglandin E synthase-1, is the mechanism for interleukin-1-β-induced prostaglandin E$_2$ production and inhibition of insulin secretion in pancreatic islets. *J. Biol. Chem.* **2012**, *287*, 32246–32253. [CrossRef]

21. Tran, P.O.T.; Gleason, C.E.; Poitout, V.; Robertson, R.P. Prostaglandin E$_2$ mediates inhibition of insulin secretion by interleukin-1β. *J. Biol. Chem.* **1999**, *274*, 31245–31248. [CrossRef] [PubMed]

22. Tran, P.O.T.; Gleason, C.E.; Paul Robertson, R. Inhibition of interleukin-1β-induced COX-2 and EP3 gene expression by sodium salicylate enhances pancreatic islet β-cell function. *Diabetes* **2002**, *51*, 1772–1778. [CrossRef] [PubMed]

23. Maedler, K.; Sergeev, P.; Ris, F.; Oberholzer, J.; Joller-Jemelka, H.I.; Spinas, G.A.; Kaiser, N.; Halban, P.A.; Donath, M.Y. Glucose-induced β cell production of IL-1β contributes to glucotoxicity in human pancreatic islets. *J. Clin. Investig.* **2002**, *110*, 851–860. [CrossRef] [PubMed]

24. Nieuwdorp, M.; Gilijamse, P.W.; Pai, N.; Kaplan, L.M. Role of the microbiome in energy regulation and metabolism. *Gastroenterology* **2014**, *146*, 1525–1533. [CrossRef]

25. Larsen, N.; Vogensen, F.K.; Van Den Berg, F.W.J.; Sandris Nielsen, D.; Andreasen, A.S.; Pedersen, B.K.; Al-Soud, W.A.; Sørensen, S.J.; Hansen, L.H.; Jakobsen, M. Gut Microbiota in Human Adults with Type 2 Diabetes Differs from Non-Diabetic Adults. *PLoS ONE* **2010**, *5*, e9085. [CrossRef]

26. Rizzatti, G.; Lopetuso, L.R.; Gibiino, G.; Binda, C.; Gasbarrini, A. Proteobacteria: A common factor in human diseases. *Biomed. Res. Int.* **2017**, *2017*. [CrossRef]

27. Singh, R.K.; Chang, H.W.; Yan, D.; Lee, K.M.; Ucmak, D.; Wong, K.; Abrouk, M.; Farahnik, B.; Nakamura, M.; Zhu, T.H.; et al. Influence of diet on the gut microbiome and implications for human health. *J. Transl. Med.* **2017**, *15*. [CrossRef]

28. Yamane, S.; Inagaki, N. Regulation of glucagon-like peptide-1 sensitivity by gut microbiota dysbiosis. *J. Diabetes Investig.* **2018**, *9*, 262–264. [CrossRef]

29. Baggio, L.L.; Drucker, D.J. Biology of Incretins: GLP-1 and GIP. *Gastroenterology* **2007**, *132*, 2131–2157. [CrossRef]

30. Correia, M.L.G.; Haynes, W.G. A role for plasminogen activator inhibitor-1 in obesity: From pie to PAI? *Arterioscler. Thromb. Vasc. Biol.* **2006**, *26*, 2183–2185. [CrossRef]

31. Martin, A.M.; Sun, E.W.; Rogers, G.B.; Keating, D.J. The influence of the gut microbiome on host metabolism through the regulation of gut hormone release. *Front. Physiol.* **2019**, *10*. [CrossRef] [PubMed]

32. Beier, J.I.; Guo, L.; Von Montfort, C.; Kaiser, J.P.; Joshi-Barve, S.; Arteel, G.E. New role of resistin in lipopolysaccharide-induced liver damage in mice. *J. Pharmacol. Exp. Ther.* **2008**, *325*, 801–808. [CrossRef] [PubMed]

33. Kenny, S.; Gamble, J.; Lyons, S.; Vlatković, N.; Dimaline, R.; Varro, A.; Dockray, G.J. Gastric expression of plasminogen activator inhibitor (PAI)-1 is associated with hyperphagia and obesity in mice. *Endocrinology* **2013**, *154*, 718–726. [CrossRef] [PubMed]

34. Byberg, L.; Smedman, A.; Vessby, B.; Lithell, H. Plasminogen activator inhibitor-1 and relations to fatty acid composition in the diet and in serum cholesterol esters. *Arterioscler. Thromb. Vasc. Biol.* **2001**, *21*, 2086–2092. [CrossRef]

35. Cesari, M.; Pahor, M.; Incalzi, R.A. Plasminogen activator inhibitor-1 (PAI-1): A key factor linking fibrinolysis and age-related subclinical and clinical conditions. *Cardiovasc. Ther.* **2010**, *28*. [CrossRef]

36. Kenny, S.; Duval, C.; Sammut, S.J.; Steele, I.; Pritchard, D.M.; Atherton, J.C.; Argent, R.H.; Dimaline, R.; Dockray, G.J.; Varro, A. Increased expression of the urokinase plasminogen activator system by *Helicobacter pylori* in gastric epithelial cells. *Am. J. Physiol. Gastrointest. Liver Physiol.* **2008**, *295*. [CrossRef]

37. Cani, P.D.; Amar, J.; Iglesias, M.A.; Poggi, M.; Knauf, C.; Bastelica, D.; Neyrinck, A.M.; Fava, F.; Tuohy, K.M.; Chabo, C.; et al. Metabolic endotoxemia initiates obesity and insulin resistance. *Diabetes* **2007**, *56*, 1761–1772. [CrossRef]

38. Cani, P.D.; Bibiloni, R.; Knauf, C.; Waget, A.; Neyrinck, A.M.; Delzenne, N.M.; Burcelin, R. Changes in gut microbiota control metabolic endotoxemia-induced inflammation in high-fat diet-induced obesity and diabetes in mice. *Diabetes* **2008**, *57*, 1470–1481. [CrossRef]

39. Amyot, J.; Semache, M.; Ferdaoussi, M.; Fontés, G.; Poitout, V. Lipopolysaccharides impair insulin gene expression in isolated islets of langerhans via toll-like receptor-4 and nf-κb signalling. *PLoS ONE* **2012**, *7*, e36200. [CrossRef]

40. Lu, S.C.; Shieh, W.Y.; Chen, C.Y.; Hsu, S.C.; Chen, H.L. Lipopolysaccharide increases resistin gene expression in vivo and in vitro. *FEBS Lett.* **2002**, *530*, 158–162. [CrossRef]

41. Poreba, M.; Rostoff, P.; Siniarski, A.; Mostowik, M.; Golebiowska-Wiatrak, R.; Nessler, J.; Undas, A.; Gajos, G. Relationship between polyunsaturated fatty acid composition in serum phospholipids, systemic low-grade inflammation, and glycemic control in patients with type 2 diabetes and atherosclerotic cardiovascular disease. *Cardiovasc. Diabetol.* **2018**, *17*. [CrossRef] [PubMed]

42. Kimple, M.E.; Moss, J.B.; Brar, H.K.; Rosa, T.C.; Truchan, N.A.; Pasker, R.L.; Newgard, C.B.; Casey, P.J. Deletion of Gαz protein protects against diet-induced glucose intolerance via expansion of β-cell mass. *J. Biol. Chem.* **2012**, *287*, 20344–20355. [CrossRef] [PubMed]

43. Neuman, J.C.; Truchan, N.A.; Joseph, J.W.; Kimple, M.E. A method for mouse pancreatic islet isolation and intracellular cAMP determination. *J. Vis. Exp.* **2014**. [CrossRef] [PubMed]

44. Truchan, N.A.; Brar, H.K.; Gallagher, S.J.; Neuman, J.C.; Kimple, M.E. A single-islet microplate assay to measure mouse and human islet insulin secretion. *Islets* **2015**, *7*, 1–10. [CrossRef]

45. Brill, A.L.; Wisinski, J.A.; Cadena, M.T.; Thompson, M.F.; Fenske, R.J.; Brar, H.K.; Schaid, M.D.; Pasker, R.L.; Kimple, M.E. Synergy between Gαz deficiency and GLP-1 analog treatment in preserving functional β-cell mass in experimental diabetes. *Mol. Endocrinol.* **2016**, *30*, 543–556. [CrossRef]

46. Pak, H.H.; Cummings, N.E.; Green, C.L.; Brinkman, J.A.; Yu, D.; Tomasiewicz, J.L.; Yang, S.E.; Boyle, C.; Konon, E.N.; Ong, I.M.; et al. The Metabolic Response to a Low Amino Acid Diet is Independent of Diet-Induced Shifts in the Composition of the Gut Microbiome. *Sci. Rep.* **2019**, *9*. [CrossRef]
47. Caporaso, J.G.; Bittinger, K.; Bushman, F.D.; Desantis, T.Z.; Andersen, G.L.; Knight, R. PyNAST: A flexible tool for aligning sequences to a template alignment. *Bioinformatics* **2010**, *26*, 266–267. [CrossRef]
48. Chong, J.; Soufan, O.; Li, C.; Caraus, I.; Li, S.; Bourque, G.; Wishart, D.S.; Xia, J. MetaboAnalyst 4.0: Towards more transparent and integrative metabolomics analysis. *Nucleic Acids Res.* **2018**, *46*, W486–W494. [CrossRef]
49. Perez-Riverol, Y.; Csordas, A.; Bai, J.; Bernal-Llinares, M.; Hewapathirana, S.; Kundu, D.J.; Inuganti, A.; Griss, J.; Mayer, G.; Eisenacher, M.; et al. The PRIDE database and related tools and resources in 2019: Improving support for quantification data. *Nucleic Acids Res.* **2019**, *47*, D442–D450. [CrossRef]

 *metabolites*

Article

Single-Cell Transcriptional Profiling of Mouse Islets Following Short-Term Obesogenic Dietary Intervention

Annie R. Piñeros [1], Hongyu Gao [2], Wenting Wu [1,2], Yunlong Liu [2], Sarah A. Tersey [3] and Raghavendra G. Mirmira [3,*]

[1]   Department of Pediatrics, Indiana University School of Medicine, Indianapolis, IN 46202, USA; apineros@iupui.edu (A.R.P.); wuwent@iu.edu (W.W.)

[2]   Department of Medical and Molecular Genetics, Indiana University School of Medicine, Indianapolis, IN 46202, USA; hongao@iu.edu (H.G.); yunliu@iu.edu (Y.L.)

[3]   Kolver Diabetes Center and Department of Medicine, The University of Chicago, Chicago, IL 60637, USA; stersey@uchicago.edu

*   Correspondence: mirmira@uchicago.edu

Received: 16 November 2020; Accepted: 17 December 2020; Published: 18 December 2020

Abstract: Obesity is closely associated with adipose tissue inflammation and insulin resistance. Dysglycemia and type 2 diabetes results when islet β cells fail to maintain appropriate insulin secretion in the face of insulin resistance. To clarify the early transcriptional events leading to β-cell failure in the setting of obesity, we fed male C57BL/6J mice an obesogenic, high-fat diet (60% kcal from fat) or a control diet (10% kcal from fat) for one week, and islets from these mice (from four high-fat- and three control-fed mice) were subjected to single-cell RNA sequencing (sc-RNAseq) analysis. Islet endocrine cell types (α cells, β cells, δ cells, PP cells) and other resident cell types (macrophages, T cells) were annotated by transcript profiles and visualized using Uniform Manifold Approximation and Projection for Dimension Reduction (UMAP) plots. UMAP analysis revealed distinct cell clusters (11 for β cells, 5 for α cells, 3 for δ cells, PP cells, ductal cells, endothelial cells), emphasizing the heterogeneity of cell populations in the islet. Collectively, the clusters containing the majority of β cells showed the fewest gene expression changes, whereas clusters harboring the minority of β cells showed the most changes. We identified that distinct β-cell clusters downregulate genes associated with the endoplasmic reticulum stress response and upregulate genes associated with insulin secretion, whereas others upregulate genes that impair insulin secretion, cell proliferation, and cell survival. Moreover, all β-cell clusters negatively regulate genes associated with immune response activation. Glucagon-producing α cells exhibited patterns similar to β cells but, again, in clusters containing the minority of α cells. Our data indicate that an early transcriptional response in islets to an obesogenic diet reflects an attempt by distinct populations of β cells to augment or impair cellular function and/or reduce inflammatory responses as possible harbingers of ensuing insulin resistance.

Keywords: islet; obesity; insulin; transcriptomics

---

1. Introduction

An analysis of the National Health and Nutrition Examination Survey database revealed that the crude prevalence of prediabetes and diabetes in the US exceeds 50% [1]. Cardiovascular consequences, including stroke, myocardial infarction, and mortality, increase even as blood glucose rises in the prediabetic phase [2]. Type 2 diabetes (T2D) has been increasing in incidence in the USA in a manner that has tracked closely with the increasing prevalence of abdominal obesity [3]. Notably, across the spectrum from obesity to T2D, insulin resistance appears to be a common feature, and as such, insulin-producing islet β cells are considered central determinants in the transition from normoglycemia to dysglycemia. The gradual or sudden failure to secrete sufficient insulin heralds rising glucose levels [4]. Precisely how and why β cells fail have been the subjects of intense investigation in recent years. One point of view posits that the increased demand for insulin secretion in the setting of insulin resistance overwhelms the capacity of β cells to produce insulin, largely as a result of the incapacity of the endoplasmic reticulum to accommodate protein throughput [5]. Another perspective suggests that excessive circulating proinflammatory cytokines or free fatty acids trigger intracellular signaling cascades that lead to β-cell inflammation, oxidative stress, and possibly cell death [6]. Finally, recent data have suggested the provocative hypothesis that secretory failure and apoptosis may only represent minor components, and the dedifferentiation of β cells to a precursor-like state effectively leads to reduced functional β-cell mass [7]. Regardless of the primary pathology, a better understanding of the molecular processes occurring in the β cell during early obesity and dysglycemia might permit targeted therapeutic interventions that allow for disease modification.

With the advent of single-cell RNA sequencing (sc-RNAseq) technologies [8], important new insights have been emerging about the nature of cells in the pancreatic islet. For example, it was shown that α cells and β cells of the adult human islet exist in multiple subpopulations, and some of those from adults with T2D exhibit features similar to those of children, supporting the dedifferentiation hypothesis [9]. Data of this sort both illuminate the potential to extract information that would be masked using bulk sequencing strategies and emphasize the heterogeneity that exists within the islet endocrine cell pool and how specific populations might have broader implications for emerging metabolic phenotypes in an organism. In light of these and other findings, in this study, we asked if alterations in islet cell gene expression patterns occur in the earliest phases of obesity and dysglycemia in a mouse model of evolving obesity and specifically how these changes might have the ability to portend future functional alterations that give rise to overt T2D. Our data show that early changes in β-cell gene expression reflect an ability of distinct β-cell populations to respond to the metabolic challenges during high-fat-diet (HFD) feeding.

2. Results

2.1. Identification of Islet Cell Type Clusters by sc-RNAseq

Previous studies from our group demonstrated that a one-week period of HFD (60% kcal from fat) feeding compared to a low-fat diet (LFD, 10% kcal from fat) in mice led to a significant increase in mRNA translation, mTOR pathway activation, and cellular proliferation in pancreatic islets [10,11]. However, these prior studies do not provide the granularity to assess which of the specific endocrine cellular populations account for these responses. To gain more insight into the molecular pathways and the individual changes happening at the single-cell level, we performed sc-RNAseq from a collection of dissociated islets (180–200 islets/mouse) isolated from a total of seven male C57BL/6J mice fed for one week with either a HFD (60% kcal from fat, n = 4) or a control LFD (10% kcal from fat, n = 3). Each mouse sample was processed separately. Our analysis of sc-RNAseq allowed the annotation of islet cell types into different clusters based on the expression of key identifying genes, depicted in the Uniform Manifold Approximation and Projection for Dimension Reduction (UMAP) plots in Figure 1a and Supplementary Figure S1. We identified 26 cell clusters representing 9 distinct cell types based on the gene expression of canonical markers and, as expected, the major populations identified were β cells (Ins1+, Ins2+, Mafa+), α cells (Gcg+, Ttr+, Irx2+), and δ cells (Sst+, Ghsr+, Rbp4+), with smaller populations of other cell types, including PP cells (Ppy+), ductal (Krt19+, Hnf1b+), endothelial cells (Plvap+, Esm1+), immune cells as macrophages (Adgre1+, Lyz2+), T cells (Trbc2+, Cd3g+, Cd4+), and B cells (Cd19+, Igkc+, Ighm+) (Figure 1a). Notably, no differences were observed in the percentage of cells identified in any given cluster in HFD-fed compared to control LFD-fed mice (Figure 1b–f). Furthermore, we found that the percentage of total α, β, and δ cells or other types of cells identified with sc-RNAseq does not change substantially within replicates and between the LFD and HFD groups (Supplementary Table S1). These results demonstrate the reproducibility of our data at a single-cell level.

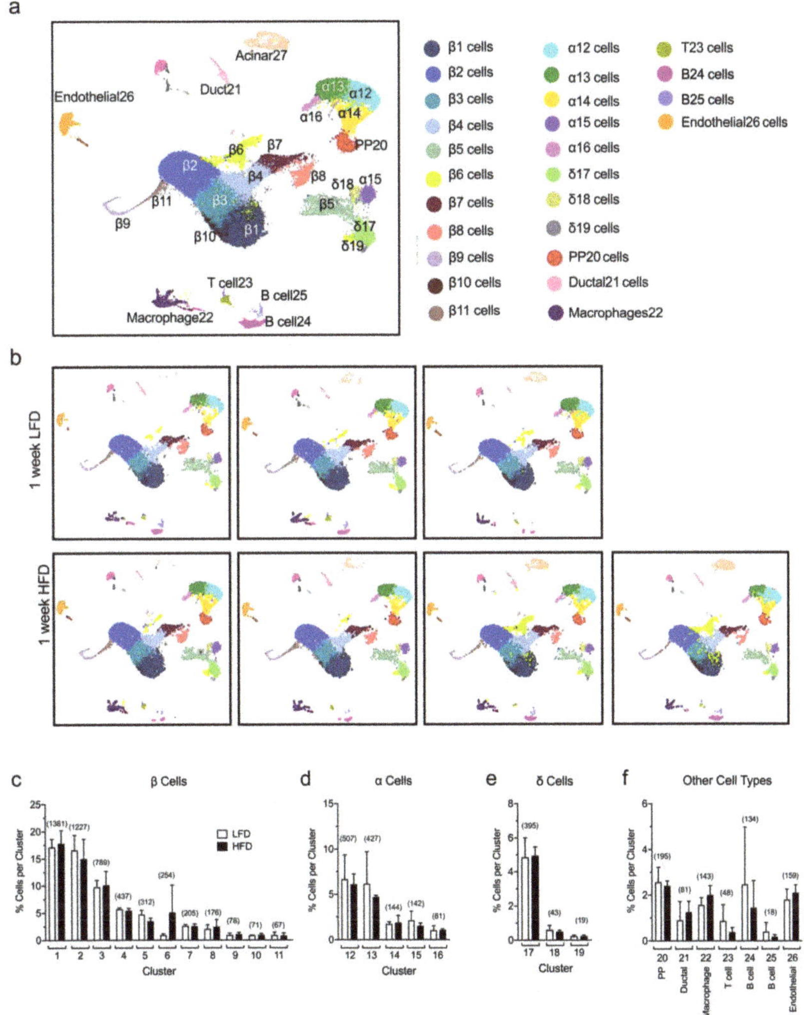

Figure 1. Identification of islet cell type clusters by single-cell RNA sequencing. Single cells were obtained from dissociated islets from male *C57BL/6J* mice fed for one week with either a high-fat diet (HFD, 60% kcal from fat, *n* = 4) or a control low-fat diet (LFD, 10% kcal from fat, *n* = 3) and used to perform single-cell RNA sequencing. (**a**) Annotation of islet cell types into different clusters based on the expression of key identifying genes, depicted in the Uniform Manifold Approximation and Projection for Dimension Reduction (UMAP) plots of merged sc-RNAseq profiles from LFD and HFD mice; (**b**) UMAP of single-cell RNA sequencing profiles from islets of individual mice fed a HFD or a LFD, as indicated; In (**c–f**), the percentage of β cells (**c**), α cells (**d**), δ cells (**e**), and other cell types (**f**) identified per cluster relative to the total number of cells sequenced is shown. Data are mean ± SEM, *n* = 3 for LFD and *n*= 4 for HFD. The number in the *parentheses* above each bar indicates the average number of cells per cluster.

2.2. Single-Cell RNA Sequencing Analysis Reveals Greatest Gene Expression Changes in Minor β Cell Clusters Following Short-Term HFD Feeding

Next, we examined the gene expression profiles in β cells from mice fed either a HFD or LFD for one week. UMAP analysis identified a total of 11 distinct β-cell clusters (β1–β11, see Figures 1a and 2a). Based on the proportion of cells per cluster (Figure 1c), we identified three major clusters of β cells (β1–β3) and eight minor clusters (β4–β11). Differential gene expression between HFD and LFD were interrogated, and statistical significance was determined by using edgeR on the integrated single-cell data obtained by the R package Seurat (see Section 4). It is notable that the major clusters of β cells (β1–β3) showed minimal change in gene expression patterns (Supplementary Figure S2), whereas the greatest changes were observed in the minor clusters (most notably β5, β7, β8, β10, β11) (Figure 2b). These findings suggest that minor β-cell clusters drive the earliest responses to HFD, and further emphasize how bulk RNA sequencing might miss these sentinel changes. In these minor clusters, the most notable gene expression changes reflect upon hormone secretion and intracellular inflammatory pathways. Clusters β5, β7, β8, β10, and β11 demonstrated significant *decreases* in genes that promote insulin secretion (*Atf6, Meg3, Herpud1*), β cell survival (*Sox4, Tnfaip3*), calcium signaling (*Robo2*), and activation of immune response (*Nfkb1, Il1r1, Cxcl10, Ifngr2*) (Figure 2b), with increases in genes that block insulin release (*Mt2*) and promote apoptosis (*Mif*) [12–17]. Furthermore, we observed an increased expression of *Mafa*, a gene that encodes a transactivator of the *Ins1/2* genes, in this group of clusters following HFD feeding.

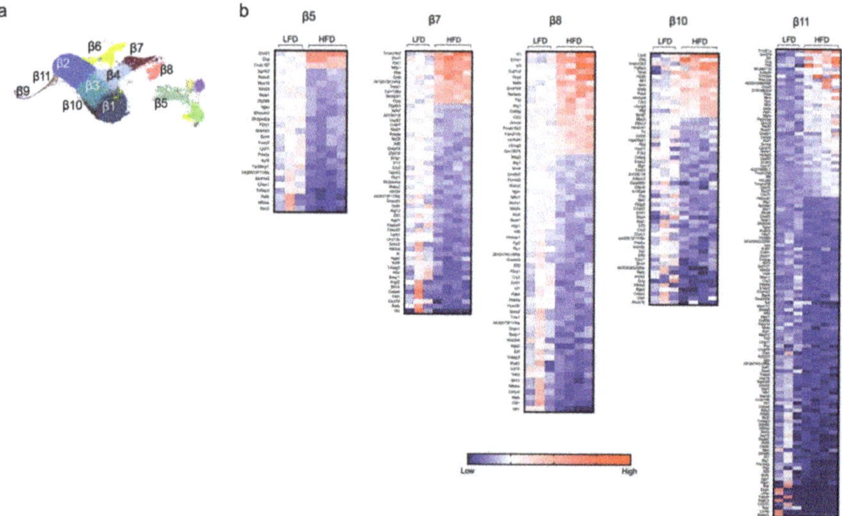

Figure 2. Identification of differentially expressed genes of the minor β-cell clusters. B-cell clusters were identified from dissociated islets from male *C57BL/6J* mice fed for one week with either a high-fat diet (HFD, 60% kcal from fat) or a control low-fat diet (LFD, 10% kcal from fat). (**a**) Representative UMAP plot of β-cell clusters identified by single-cell RNA sequencing; (**b**) heatmaps of the minor β-cell clusters of genes significantly differentially expressed ($p < 0.05$) in the β-cell clusters β5, β7, β8, β10, and β11; genes are ordered from most positive to most negative fold-change.

Among the β-cell clusters that showed minimal gene expression changes between HFD and LFD, four of them (β1, β2, β6, and β9) showed downregulation of genes for the endoplasmic reticulum stress response (*Chac1*) and activation (*Nfkbia, Cxcl10*) of inflammatory processes (Supplementary Figure S2) [18,19]. Clusters β3 and β4 showed an increase in expression of genes associated with insulin secretion (*Dbp*) and a complementary decrease in expression of genes encoding a suppressor of insulin signaling (*Pdea5*) and inflammatory pathways (*Nfkbia, Tnfaip3*) (Supplementary Figure S2) [20,21].

Collectively, these gene expression changes provide a picture of the response of β cells to HFD feeding—namely, that specific, larger clusters of β cells demonstrate augmented insulin secretory capacity and reduction of ER stress, whereas smaller clusters exhibit a reduced insulin secretion capacity and survival. It remains unclear if these findings signify an early response of small clusters, which will be later reflected in larger clusters as the impact of HFD is prolonged, or if they reflect what will be a persistent, and perhaps competing, heterogenous response among β-cell subtypes whose function on-balance determines T2D outcomes. In this regard, our prior studies of HFD feeding in male mice [22] were suggestive of recurring patterns of β-cell loss, wherein it was suggested that different subpopulations of β cells might have shown evolving susceptibility to death or dedifferentiation as HFD feeding progressed.

Using the same diet that we have used in the present work, other studies from our group have demonstrated that although one week of HFD feeding results in only minor impairments of glucose tolerance, longer-term HFD feeding impairs both glucose tolerance and insulin secretion [10,11,22]. Therefore, further studies would be necessary to know if β-cell dysfunction in the smaller clusters drives hyperglycemia and impairs insulin secretion during obesity. Several studies have also demonstrated that proinflammatory signals, cellular stress, and genetic components contribute to T2D development [23–26]. Here, we found that despite the HFD feeding, β-cell clusters overall tend to decrease the expression of proinflammatory genes, suggesting that cellular stress and β-cell dysfunction are the initial triggers that propagate the inflammatory response in the HFD feeding model. However, more studies will be necessary to clarify if the events of cellular stress and β-cell dysfunction identified in the smaller clusters of β cells propagate cellular stress in other larger clusters.

2.3. Gene Pathway Analysis Reveals Molecular Responses Related to Inflammation/Immunity and Oxidative Stress in β Cells Following HFD Feeding

Whereas the preceding studies focused on specific islet-cell gene expression changes between HFD- and LFD-fed mice, they do not provide an unbiased context for how these alterations affect functional signaling pathways in the cell. Therefore, we performed gene ontology gene set enrichment analysis (GO GSEA) to evaluate the signaling pathways specifically altered by HFD feeding in different β-cell clusters. We identified three general pathway responses exemplified by clusters β1, β4, and β7 (Figure 3). β-cell clusters with minimal gene expression changes in response to HFD had an overrepresentation of gene pathways in the response to cellular metabolism, endoplasmic reticulum (ER) stress, and oxidation–reduction process (Figure 3). A second pattern, seen largely among clusters β3 and β4 (Figure 3), showed an overrepresentation of genes associated with cell differentiation/development, immune response, and the negative regulation of apoptosis. Finally, a third group, exemplified by cluster β7 and exhibited the greatest gene expression changes in response to a HFD, showed changes in gene pathways that modulate immune/inflammatory response and cellular stress (Figure 3). Collectively, these pathway analyses are suggestive of β-cell adaptations to HFD feeding in which inflammation/immune cascades are affected and in which changes to cellular redox and metabolite utilization prevail. Our findings are consistent with the observations in the literature that free fatty acids present in HFDs impose oxidative stress [6] and enhance signaling pathways linked to cytokines and inflammation [27].

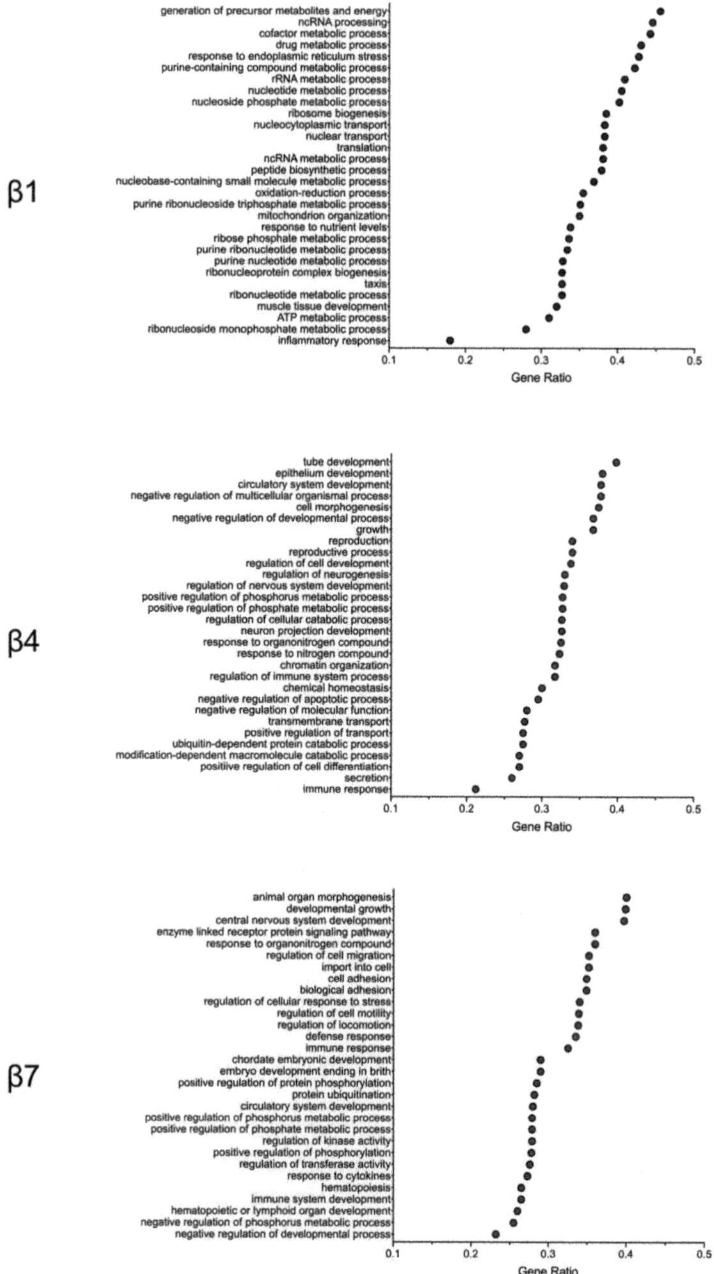

Figure 3. Altered signaling pathways of the minor β-cell clusters. Gene ontology gene set enrichment analysis (GO GSEA) was performed after single-cell RNA sequencing to evaluate the signaling pathways specifically altered by high-fat-diet feeding vs. low-fat-diet feeding in different β-cell clusters. Shown are the GO GSEA analyses of β1, β4, and β7 cell clusters.

Short-Term HFD Feeding Reveals Heterogeneity in α Cell Responses

Islet α cells produce glucagon, a major insulin counterregulatory hormone. The dysregulated hypersecretion of glucagon contributes to dysglycemia in obesity and T2D [28]. Bulk RNA-seq studies from long-term HFD-fed mice previously demonstrated that α cells display remarkably minor changes in transcriptome profile [29], but such studies might belie greater changes in gene expression patterns in specific subsets of α cells. In this study, sc-RNAseq identified minor distinct clusters of α cells (designated α12–α16, Figure 4a), the populations of which were not altered upon HFD feeding of mice for one week (Figure 1d). Differential gene expression analysis of these clusters was then performed. The cluster containing the largest pool of α cells (α12) did not show statistically significant changes in gene expression; however, clusters α13, α14, α15, and α16 showed multiple differentially expressed genes in HFD-fed mice compared to LFD-fed mice (Figure 4b). Notably, we observed that HFD feeding led to a decrease in genes associated with inflammation (*Il1r, Cxcl1, and Nfkbia*) in all four α-cell clusters (Figure 4b). Only cluster α13 showed changes in hormone expression, with an increase in Ppy after HFD feeding, indicating that, in general, short-term HFD does not lead to the misexpression of hormone-encoding genes. Other general observations include an increase in the expression of genes involved in cell survival and proliferation (*Upk3a*) and cellular stress (*Hspa1a* and *Hspa1b*) in clusters α14 and α16 [30,31], the enhanced expression of genes related to endocrine progenitors (*Neurog3*), and an oxidative stress/ER stress (*Ero1b*) decrease in cluster α15 [32,33]. Collectively, these data indicate that, like β cells, the early gene expression changes in α cells (a) are dependent on the specific cluster and thereby exhibit cluster-dependent heterogeneity, and (b) might be missed by bulk sequencing approaches since a major cluster of α cells exhibits no significant changes.

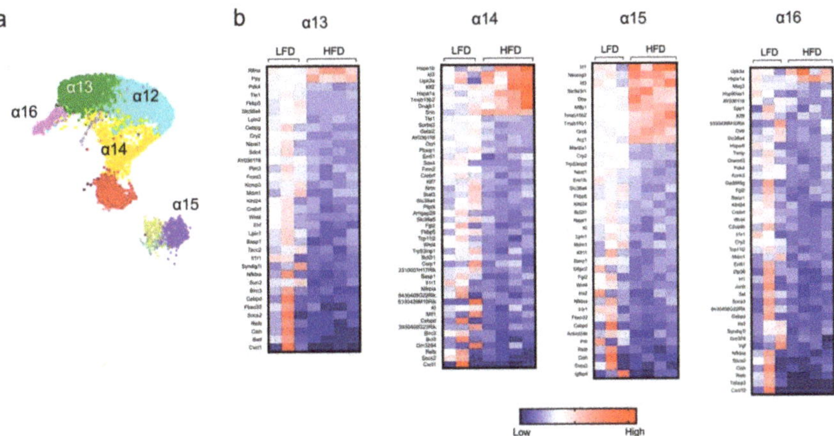

Figure 4. Transcriptome profile in α-cell clusters identified by single-cell RNA sequencing. α-cell clusters were identified following single-cell RNA sequencing from male *C57BL/6J* mice fed for one week with either a high-fat diet (HFD, 60% kcal from fat) or a control low-fat diet (LFD, 10% kcal from fat). (**a**) Representative UMAP plot of α-cell clusters; (**b**) heatmaps depicting genes differentially regulated ($p < 0.05$) in α-cell clusters; genes are ordered from most positive to most negative fold-change.

3. Discussion

To our knowledge, no studies to date have assessed the heterogeneity of individual islet cell responses to HFD feeding in mice. The use of sc-RNAseq technologies allows us to employ clustering analysis to segregate populations of cells that exhibit similar gene expression patterns and identify the heterogeneity of cellular responses. In this study, we were interested in the early responses of islet endocrine cells to HFD feeding. Prior studies have shown that one week of 60% kcal from fat a HFD results in mild glucose intolerance in male C57BL/6 mice [10,34], suggesting that β-cell failure in the setting of HFD-induced insulin resistance may occur early during obesity. Our results from sc-RNAseq analysis show that the minor clusters of β cells (β5, β7, β8, β10, and β11) exhibit an impairment in β cell function and survival, whereas some of the major clusters of β cells (β3 and β4) demonstrate transcriptomic changes consistent with a compensatory response; however, others (β1 and β2) still exhibit no statistically significant change. Our findings, therefore, suggest that the failure of a majority of β cells to compensate for prevailing insulin needs might portend the eventual development of hyperglycemia/T2D. Further studies that include analysis of several time points would provide more insight into whether the defective early response found in the minor β cells clusters triggers T2D or if it is just the result of a compensatory response to an obesogenic diet. Similarly, the robustness of a large proportion of α cells, which exhibit gene expression changes that promote proliferation and survival, might permit the more robust production of glucagon that exacerbates dysglycemia. Our studies support the concept that the balance of different populations of cell types in the islet (those more capable of responding to prevailing stress vs. those less capable) might account for the eventual risk for the development of dysglycemia/T2D in the setting of HFDs.

Several limitations should be noted in our study. First, our study was limited to an early period (one week) following obesogenic diet exposure. Therefore, we cannot know if/how gene expression responses change with time. Do the responsive β-cell clusters remain robustly responsive in their gene expression patterns with more prolonged feeding? Do the less responsive clusters become more responsive with time? Are specific clusters more susceptible to apoptosis or dedifferentiation over time? Does the islet isolation process influence mRNA expression? In future studies, applying our annotation and clustering analyses should allow us to address these questions using timed feeding cohorts. Second, our study was limited to male C57BL/6J mice, which are known to be the most susceptible to the deleterious effects of HFD feeding. The use of mice with more robust islet responsiveness, such as the C57BLKs/J strain [35] or female C57BL/6J mice [36], is likely to provide more insight into the nature and number of β-cell clusters that are responsive to the demand for insulin production. Third, the relevance of our study to human obesity and T2D remains unclear because of species differences and the added genetic heterogeneity in humans that compounds the inherent heterogeneity of the islet cell types. Fourth, although our data are compared to controls, because the procedure of islet isolation is a stressful process, we cannot rule out that it could lead to more exaggerated changes in gene expression profile that may impact our findings. Finally, we should note that sc-RNAseq may underestimate the nature of gene expression changes since the depth of sequencing by single-cell technologies is considerably lower than that seen in bulk sequencing technologies. Thus, our findings regarding the minimal gene expression changes seen in the majority of β-cell clusters may not fully report actual changes occurring in these cells. Nevertheless, our studies provide new biologic insight into the early changes in gene expression that occurs in islet cell subsets in the early phase of obesogenic diets.

4. Materials and Methods

4.1. Animal Studies

Male C57BL/6J mice were purchased from Jackson Laboratories at 7 weeks of age. All mice were kept on a standard 12 h–12 h light–dark cycle with ad libitum access to chow and water. Mice were maintained at the Indiana University vivarium according to protocols approved by the Institutional Animal Care and Use Committee. At 8 weeks of age, animals were fed either a LFD (10% kcal from fat; Research Diets D12450B) or a HFD (60% kcal from fat; Research Diets D12492) for one week.

4.2. Islet Isolation

Mouse islets were isolated from collagenase-perfused pancreata, as previously described [37]. After isolation, mouse islets (approximately 180–200 islets) were handpicked and digested with Accutase (EMD Millipore Corporation, Temecula, CA, USA) containing 2 U/mL of DNAse for 5 min at 37 °C sob agitation (1000 rpm). Digested cell islets were washed several times with PBS+2%FBS to eliminate DNAse and then filtered using a cell strainer (40 μm). Single-cell suspensions and samples with more than 90% viability were used for sc-RNAseq.

4.3. Preparation of Single Cell 3′ RNA-seq Library

Single-cell 3′ RNA-seq experiments were conducted using the Chromium single-cell system (10 x Genomics, Inc., Pleasanton, CA, USA) and Illumina sequencers at the Center for Medical Genetics (CMG) of the Indiana University School of Medicine. Each cell suspension, a collection of 180–200 islets isolated from an individual mouse, was first inspected under a microscope for cell number, cell viability, and cell size. For the initial cell suspension with low viability, the single-cell preparation was further processed, which included centrifugation, resuspension, and filtration to remove cell debris, dead cells, and cell aggregates. Single-cell capture and library preparation were performed according to the Chromium Single Cell 3′ Reagent Kits V3 User Guide (10x Genomics PN-1000075, PN-1000073, PN-120262). An appropriate number of cells were loaded on a multiple-channel microfluidics chip of the Chromium Single Cell Instrument (10x Genomics) with a targeted cell recovery of 8000 to 10,000. Single-cell gel beads in an emulsion containing barcoded oligonucleotides and reverse-transcriptase reagents were generated with the V3 single-cell reagent kit (10x Genomics). Following cell capture and cell lysis, cDNA was synthesized and amplified. Illumina sequencing libraries were then prepared with the amplified cDNA. The resulting libraries were assessed with Agilent TapeStation. The final libraries were sequenced using a custom program on Illumina NovaSeq 6000. About 550 million read pairs were generated for each sample, with 28 bp of cell barcode and unique molecular identifier reads and 91 bp RNA reads.

4.4. Analysis of sc-RNAseq Sequence Data

CellRanger 3.0.2 (http://support.10xgenomics.com/) was utilized to process the raw sequence data generated. Briefly, CellRanger used bcl2fastq (https://support.illumina.com/) to demultiplex raw base sequence calls generated from the sequencer into sample-specific FASTQ files. The FASTQ files were then aligned to the mouse reference genome mm10 with RNA-seq aligner STAR. The aligned reads were traced back to individual cells, and the gene expression level of individual genes was quantified based on the number of UMIs detected in each cell.

The filtered gene–cell barcode matrices generated from CellRanger were used for further analysis with the R package Seurat (Seurat development version 3.0.0.9200) [38,39] with Rstudio version 1.1.453 and R version 3.5.1. Quality control (QC) of the data was implemented as the first step in our analysis. We filtered out genes that were detected in less than five cells and cells with less than 200 genes. To further exclude low-quality cells in downstream analysis, we used the Outlier function from the R package scatter [40] together with a visual inspection of the distributions of the number of genes, UMIs, and mitochondrial gene content. Cells with an extremely high or low number of detected genes/UMIs were excluded. In addition, cells with a high percentage of mitochondrial reads were also filtered out. After removing likely multiplets and low-quality cells, the gene expression levels for each cell were normalized with the NormalizeData function in Seurat. Highly variable genes were subsequently identified.

To integrate the single-cell data of the HFD and LFD samples, functions FindIntegrationAnchors and IntegrateData from Seurat were applied. The integrated data were scaled, and PCA was performed. Clusters were identified with the Seurat functions FindNeighbors and FindClusters. The FindConservedMarkers function was subsequently used to identify cell cluster marker genes. Cell cluster identities were manually defined with the cluster-specific marker genes or known marker genes. The cell clusters were visualized using the t-Distributed Stochastic Neighbor Embedding (t-SNE) plots and Uniform Manifold Approximation and Projection (UMAP) plots.

To investigate cell cluster/type-specific responses triggered by different diet treatments, a pseudobulk count approach was applied. The counts of each gene of all cells within a cell cluster for each biological replicate were aggregated. In this way, the gene–cell count matrix was transformed into a gene–sample level count matrix. The R package muscat (version 1.0.1, [41]) was employed for generating pseudobulk counts and differential gene expression analysis between conditions with the same cluster. In more details, the Seurat object generated from the integrative analysis was first converted into a SingleCellExperiment object using the as.SingleCellExperiment function in Seurat. The SingleCellExperiment object was then used as input for muscat. Gene–cell count data were aggregated with the function aggregateData, and differential gene expression analysis of the pseudocounts was performed with edgeR [42]. GO gene set enrichment analysis was performed using the R package clusterProfiler [43]. The datasets generated and analyzed in the present study are available in the GEO repository (accession number: GSE162512).

Supplementary Materials: The following are available online at http://www.mdpi.com/2218-1989/10/12/513/s1, Table S1: Percentage of cells per cluster identified by sc-RNAseq, Figure S1: Genes used for identification of cell clusters, Figure S2: Identification of differentially expressed genes of the major β cell clusters.

Author Contributions: Conceptualization, A.R.P., W.W., S.A.T., and R.G.M.; methodology, A.R.P. and S.A.T.; formal analysis, H.G. and Y.L.; writing—original draft preparation, A.R.P.; writing—review and editing, A.R.P., S.A.T., R.G.M.; funding acquisition, R.G.M. All authors have read and agreed to the published version of the manuscript.

Funding: This research was funded by the National Institutes of Health, grant number R01 DK60581, R01 DK 105588, P30 DK097512, and P30 DK020595.

Acknowledgments: The authors wish to acknowledge K. Orr and X. Xuei at Indiana University for technical assistance in the studies.

Conflicts of Interest: The authors declare no conflict of interest.

References

1. Menke, A.; Casagrande, S.; Geiss, L.; Cowie, C.C. Prevalence of and Trends in Diabetes Among Adults in the United States, 1988–2012. *JAMA* **2015**, *314*, 1021–1029. [CrossRef] [PubMed]

2. Lawes, C.M.M.; Parag, V.; Bennett, D.A.; Suh, I.; Lam, T.H.; Whitlock, G.; Barzi, F.; Woodward, M. Asia Pacific Cohort Studies Collaboration Blood glucose and risk of cardiovascular disease in the Asia Pacific region. *Diabetes Care* **2004**, *27*, 2836–2842. [PubMed]

3. Caspard, H.; Jabbour, S.; Hammar, N.; Fenici, P.; Sheehan, J.J.; Kosiborod, M. Recent trends in the prevalence of type 2 diabetes and the association with abdominal obesity lead to growing health disparities in the USA: An analysis of the NHANES surveys from 1999 to 2014. *Diabetes Obes. Metab.* **2018**, *20*, 667–671. [CrossRef]

4. Tabak, A.G.; Jokela, M.; Akbaraly, T.N.; Brunner, E.J.; Kivimaki, M.; Witte, D.R. Trajectories of glycaemia, insulin sensitivity, and insulin secretion before diagnosis of type 2 diabetes: An analysis from the Whitehall II study. *Lancet* **2009**, *373*, 2215–2221. [CrossRef]

5. Eizirik, D.L.; Miani, M.; Cardozo, A.K. Signalling danger: Endoplasmic reticulum stress and the unfolded protein response in pancreatic islet inflammation. *Diabetologia* **2013**, *56*, 234–241. [CrossRef]

6. Kharroubi, I.; Ladrière, L.; Cardozo, A.K.; Dogusan, Z.; Cnop, M.; Eizirik, D.L. Free Fatty Acids and Cytokines Induce Pancreatic β-Cell Apoptosis by Different Mechanisms: Role of Nuclear Factor-κB and Endoplasmic Reticulum Stress. *Endocrinology* **2004**, *145*, 5087–5096. [CrossRef]

7. Talchai, C.; Xuan, S.; Lin, H.V.; Sussel, L.; Accili, D. Pancreatic β-Cell Dedifferentiation As Mechanism Of Diabetic β-Cell Failure. *Cell* **2012**, *150*, 1223–1234. [CrossRef]

8. Chen, G.; Ning, B.; Shi, T. Single-Cell RNA-Seq Technologies and Related Computational Data Analysis. *Front. Genet.* **2019**, *10*, 317. [CrossRef]

9. Wang, Y.J.; Schug, J.; Won, K.-J.; Liu, C.; Naji, A.; Avrahami, D.; Golson, M.L.; Kaestner, K.H. Single-Cell Transcriptomics of the Human Endocrine Pancreas. *Diabetes* **2016**, *65*, 3028–3038. [CrossRef]

10. Levasseur, E.M.; Yamada, K.; Piñeros, A.R.; Wu, W.; Syed, F.; Orr, K.S.; Anderson-Baucum, E.; Mastracci, T.L.; Maier, B.; Mosley, A.L.; et al. Hypusine biosynthesis in β cells links polyamine metabolism to facultative cellular proliferation to maintain glucose homeostasis. *Sci. Signal.* **2019**, *12*, aax0715. [CrossRef]

11. Hatanaka, M.; Maier, B.; Sims, E.K.; Templin, A.T.; Kulkarni, R.N.; Evans-Molina, C.; Mirmira, R.G. Palmitate induces mRNA translation and increases ER protein load in islet β-cells via activation of the mammalian target of rapamycin pathway. *Diabetes* **2014**, *63*, 3404–3415. [CrossRef] [PubMed]

12. Adams, M.T.; Reissaus, C.A.; Szulczewski, J.M.; Dwulet, J.M.; Lyman, M.R.; Sdao, S.M.; Nimkulrat, S.D.; Ponik, S.M.; Merrins, M.J.; Benninger, R.K.P.; et al. Islet architecture controls synchronous β cell response to glucose in the intact mouse pancreas in vivo. *bioRxiv* **2019**. [CrossRef]

13. Stojanovic, I.; Saksida, T.; Timotijevic, G.; Sandler, S.; Stosic-Grujicic, S. Macrophage migration inhibitory factor (MIF) enhances palmitic acid- and glucose-induced murine beta cell dysfunction and destruction in vitro. *Growth Factors Chur Switz.* **2012**, *30*, 385–393. [CrossRef] [PubMed]

14. Tuomi, T.; Nagorny, C.L.F.; Singh, P.; Bennet, H.; Yu, Q.; Alenkvist, I.; Isomaa, B.; Östman, B.; Söderström, J.; Pesonen, A.-K.; et al. Increased Melatonin Signaling Is a Risk Factor for Type 2 Diabetes. *Cell Metab.* **2016**, *23*, 1067–1077. [CrossRef]

15. Wang, N.; Zhu, Y.; Xie, M.; Wang, L.; Jin, F.; Li, Y.; Yuan, Q.; De, W. Long Noncoding RNA Meg3 Regulates Mafa Expression in Mouse Beta Cells by Inactivating Rad21, Smc3 or Sin3α. *Cell. Physiol. Biochem.* **2018**, *45*, 2031–2043. [CrossRef] [PubMed]

16. Wong, N.; Morahan, G.; Stathopoulos, M.; Proietto, J.; Andrikopoulos, S. A novel mechanism regulating insulin secretion involving Herpud1 in mice. *Diabetologia* **2013**, *56*, 1569–1576. [CrossRef] [PubMed]

17. Xu, E.E.; Sasaki, S.; Speckmann, T.; Nian, C.; Lynn, F.C. SOX4 Allows Facultative β-Cell Proliferation Through Repression of Cdkn1a. *Diabetes* **2017**, *66*, 2213–2219. [CrossRef]

18. Mungrue, I.N.; Pagnon, J.; Kohannim, O.; Gargalovic, P.S.; Lusis, A.J. CHAC1/MGC4504 Is a Novel Proapoptotic Component of the Unfolded Protein Response, Downstream of the ATF4-ATF3-CHOP Cascade. *J. Immunol.* **2009**, *182*, 466–476. [CrossRef]

19. Liuwantara, D.; Elliot, M.; Smith, M.W.; Yam, A.O.; Walters, S.N.; Marino, E.; McShea, A.; Grey, S.T. Nuclear Factor-κB Regulates β-Cell Death: A Critical Role for A20 in β-Cell Protection. *Diabetes* **2006**, *55*, 2491–2501. [CrossRef]

20. Liu, W.; Tian, X.; Wu, T.; Liu, L.; Guo, Y.; Wang, C. PDE5A Suppresses Proteasome Activity Leading to Insulin Resistance in C2C12 Myotubes. *Int. J. Endocrinol.* **2019**, *2019*, 3054820. [CrossRef]

21. Ohta, Y.; Taguchi, A.; Matsumura, T.; Nakabayashi, H.; Akiyama, M.; Yamamoto, K.; Fujimoto, R.; Suetomi, R.; Yanai, A.; Shinoda, K.; et al. Clock Gene Dysregulation Induced by Chronic ER Stress Disrupts β-cell Function. *EBioMedicine* **2017**, *18*, 146–156. [CrossRef] [PubMed]

22. Tersey, S.A.; Levasseur, E.M.; Syed, F.; Farb, T.B.; Orr, K.S.; Nelson, J.B.; Shaw, J.L.; Bokvist, K.; Mather, K.J.; Mirmira, R.G. Episodic β-cell death and dedifferentiation during diet-induced obesity and dysglycemia in male mice. *FASEB J. Off. Publ. Fed. Am. Soc. Exp. Biol.* **2018**, *32*, 6150–6158. [CrossRef]

23. Lawlor, N.; Khetan, S.; Ucar, D.; Stitzel, M.L. Genomics of Islet (Dys)function and Type 2 Diabetes. *Trends Genet. TIG* **2017**, *33*, 244–255. [CrossRef] [PubMed]

24. Zhang, K. Integration of ER stress, oxidative stress and the inflammatory response in health and disease. *Int. J. Clin. Exp. Med.* **2010**, *3*, 33–40.

25. O'Neill, C.M.; Lu, C.; Corbin, K.L.; Sharma, P.R.; Dula, S.B.; Carter, J.D.; Ramadan, J.W.; Xin, W.; Lee, J.K.; Nunemaker, C.S. Circulating levels of IL-1B+IL-6 cause ER stress and dysfunction in islets from prediabetic male mice. *Endocrinology* **2013**, *154*, 3077–3088. [CrossRef]

26. Ghosh, R.; Colon-Negron, K.; Papa, F.R. Endoplasmic reticulum stress, degeneration of pancreatic islet β-cells, and therapeutic modulation of the unfolded protein response in diabetes. *Mol. Metab.* **2019**, *27S*, S60–S68. [CrossRef]

27. Eguchi, K.; Manabe, I.; Oishi-Tanaka, Y.; Ohsugi, M.; Kono, N.; Ogata, F.; Yagi, N.; Ohto, U.; Kimoto, M.; Miyake, K.; et al. Saturated Fatty Acid and TLR Signaling Link β Cell Dysfunction and Islet Inflammation. *Cell Metab.* **2012**, *15*, 518–533. [CrossRef]

28. Wewer Albrechtsen, N.J.; Pedersen, J.; Galsgaard, K.D.; Winther-Sørensen, M.; Suppli, M.P.; Janah, L.; Gromada, J.; Vilstrup, H.; Knop, F.K.; Holst, J.J. The Liver–α-Cell Axis and Type 2 Diabetes. *Endocr. Rev.* **2019**, *40*, 1353–1366. [CrossRef]

29. Dusaulcy, R.; Handgraaf, S.; Visentin, F.; Howald, C.; Dermitzakis, E.T.; Philippe, J.; Gosmain, Y. High-fat diet impacts more changes in beta-cell compared to alpha-cell transcriptome. *PLoS ONE* **2019**, *14*, e0213299. [CrossRef] [PubMed]

30. Drew, B.G.; Ribas, V.; Le, J.A.; Henstridge, D.C.; Phun, J.; Zhou, Z.; Soleymani, T.; Daraei, P.; Sitz, D.; Vergnes, L.; et al. HSP72 Is a Mitochondrial Stress Sensor Critical for Parkin Action, Oxidative Metabolism, and Insulin Sensitivity in Skeletal Muscle. *Diabetes* **2014**, *63*, 1488–1505. [CrossRef] [PubMed]

31. Rosas, P.C.; Nagaraja, G.M.; Kaur, P.; Panossian, A.; Wickman, G.; Garcia, L.R.; Al-Khamis, F.A.; Asea, A.A.A. Hsp72 (HSPA1A) Prevents Human Islet Amyloid Polypeptide Aggregation and Toxicity: A New Approach for Type 2 Diabetes Treatment. *PLoS ONE* **2016**, *11*, e0149409. [CrossRef] [PubMed]

32. Awazawa, M.; Futami, T.; Sakada, M.; Kaneko, K.; Ohsugi, M.; Nakaya, K.; Terai, A.; Suzuki, R.; Koike, M.; Uchiyama, Y.; et al. Deregulation of Pancreas-Specific Oxidoreductin ERO1β in the Pathogenesis of Diabetes Mellitus. *Mol. Cell. Biol.* **2014**, *34*, 1290–1299. [CrossRef] [PubMed]

33. Wang, S.; Jensen, J.N.; Seymour, P.A.; Hsu, W.; Dor, Y.; Sander, M.; Magnuson, M.A.; Serup, P.; Gu, G. Sustained Neurog3 expression in hormone-expressing islet cells is required for endocrine maturation and function. *Proc. Natl. Acad. Sci. USA* **2009**, *106*, 9715–9720. [CrossRef] [PubMed]

34. Mosser, R.E.; Maulis, M.F.; Moullé, V.S.; Dunn, J.C.; Carboneau, B.A.; Arasi, K.; Pappan, K.; Poitout, V.; Gannon, M. High-fat diet-induced β-cell proliferation occurs prior to insulin resistance in C57Bl/6J male mice. *Am. J. Physiol. Endocrinol. Metab.* **2015**, *308*, E573–E582. [CrossRef]

35. Sims, E.K.; Hatanaka, M.; Morris, D.L.; Tersey, S.A.; Kono, T.; Chaudry, Z.Z.; Day, K.H.; Moss, D.R.; Stull, N.D.; Mirmira, R.G.; et al. Divergent compensatory responses to high-fat diet between C57BL6/J and C57BLKS/J inbred mouse strains. *Am. J. Physiol. Endocrinol. Metab.* **2013**, *305*, E1495–E1511. [CrossRef]

36. Gelineau, R.R.; Arruda, N.L.; Hicks, J.A.; Monteiro De Pina, I.; Hatzidis, A.; Seggio, J.A. The behavioral and physiological effects of high-fat diet and alcohol consumption: Sex differences in C57BL6/J mice. *Brain Behav.* **2017**, *7*, e00708. [CrossRef]

37. Stull, N.D.; Breite, A.; McCarthy, R.C.; Tersey, S.A.; Mirmira, R.G. Mouse Islet of Langerhans Isolation using a Combination of Purified Collagenase and Neutral Protease. *J. Vis. Exp.* **2012**, *67*, e4137. [CrossRef]

38. Butler, A.; Hoffman, P.; Smibert, P.; Papalexi, E.; Satija, R. Integrating single-cell transcriptomic data across different conditions, technologies, and species. *Nat. Biotechnol.* **2018**, *36*, 411–420. [CrossRef]

39. Stuart, T.; Butler, A.; Hoffman, P.; Hafemeister, C.; Papalexi, E.; Mauck, W.M.; Hao, Y.; Stoeckius, M.; Smibert, P.; Satija, R. Comprehensive Integration of Single-Cell Data. *Cell* **2019**, *177*, 1888–1902. [CrossRef]

40. McCarthy, D.J.; Campbell, K.R.; Lun, A.T.L.; Wills, Q.F. Scater: Pre-processing, quality control, normalization and visualization of single-cell RNA-seq data in R. *Bioinform. Oxf. Engl.* **2017**, *33*, 1179–1186. [CrossRef]

41. Crowell, H.L.; Soneson, C.; Germain, P.-L.; Calini, D.; Collin, L.; Raposo, C.; Malhotra, D.; Robinson, M.D. On the discovery of population-specific state transitions from multi-sample multi-condition single-cell RNA sequencing data. *bioRxiv* **2019**, 713412. [CrossRef]

42. Robinson, M.D.; McCarthy, D.J.; Smyth, G.K. edgeR: A Bioconductor package for differential expression analysis of digital gene expression data. *Bioinform. Oxf. Engl.* **2010**, *26*, 139–140. [CrossRef] [PubMed]

43. Yu, G.; Wang, L.-G.; Han, Y.; He, Q.-Y. clusterProfiler: An R Package for Comparing Biological Themes Among Gene Clusters. *OMICS J. Integr. Biol.* **2012**, *16*, 284–287. [CrossRef] [PubMed]

Publisher's Note: MDPI stays neutral with regard to jurisdictional claims in published maps and institutional affiliations.

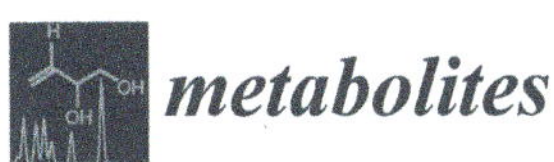 *metabolites*

Review

Good Cop, Bad Cop: The Opposing Effects of Macrophage Activation State on Maintaining or Damaging Functional β-Cell Mass

Daelin M. Jensen, Kyle V. Hendricks, Austin T. Mason and Jeffery S. Tessem *

Nutrition, Dietetics and Food Science Department, Brigham Young University, Provo, UT 84602, USA;
daelinmichael@gmail.com (D.M.J.); kylev.hendricks@gmail.com (K.V.H.);
austintaylormason@gmail.com (A.T.M.)
* Correspondence: Jeffery_tessem@byu.edu

Received: 20 October 2020; Accepted: 24 November 2020; Published: 26 November 2020

Abstract: Loss of functional β-cell mass is a hallmark of Type 1 and Type 2 Diabetes. Macrophages play an integral role in the maintenance or destruction of pancreatic β-cells. The effect of the macrophage β-cell interaction is dependent on the activation state of the macrophage. Macrophages can be activated across a spectrum, from pro-inflammatory to anti-inflammatory and tissue remodeling. The factors secreted by these differentially activated macrophages and their effect on β-cells define the effect on functional β-cell mass. In this review, the spectrum of macrophage activation is discussed, as are the positive and negative effects on β-cell survival, expansion, and function as well as the defined factors released from macrophages that impinge on functional β-cell mass.

Keywords: β-cell; macrophage; islet; cytokine; diabetes

---

1. Introduction

The prevalence of diabetes is growing. It is currently estimated that 463 million individuals are diabetic and that by the year 2045 that number will increase to 700 million [1]. While the etiologies of the two primary forms of diabetes are clearly different, Type 1 Diabetes (T1D) and Type 2 Diabetes (T2D) both result in decreased functional β-cell mass (defined as changes in β-cell survival, proliferation, and insulin secretion). T1D is characterized by autoimmune destruction of the insulin-producing pancreatic β-cells [2], and T2D is characterized by β-cell dysfunction and the ultimate loss of β-cell maturity and increased β-cell death [3].

While clearly observed with T1D, the increased presence and islet infiltration of hematopoietic cells are also observed in the pancreas of T2D patients [4,5]. Additionally, resident monocyte-derived dendritic cells and macrophages also play a critical role in β-cell homeostasis [6]. Signaling from these cell types can result in modifications in β-cell function, survival, and proliferation. The direct interaction between hematopoietic cells and β-cells plays a critical role in the maintenance of functional β-cell mass.

Resident macrophages are found in all human tissues. The entire macrophage pool in an adult human is estimated to be about 10^{10} cells [7]. Macrophages are a critical part of the innate immune response that specializes in the detection and destruction of foreign pathogens as well as the activation and recruitment of adaptive immune cells. Inflammatory macrophages have classically been considered to be detrimental to β-cell function and survival, thereby contributing to β-cell failure in both T1D and T2D. Recent findings, however, have demonstrated that anti-inflammatory macrophages play a supportive role through tissue remodeling that protects β-cells and enhances insulin secretion and replication. These contradictory effects of the macrophage on the β-cell are due to the macrophage activation state and the factors that are produced by and released from macrophages found in the pancreatic islet. In this review, the deleterious and protective effects of macrophages on the β-cell are described in the context of macrophage activation states and the factors secreted by macrophages that signal to the β-cell. Further understanding of the origins and activation pathways of tissue-resident macrophages is fundamental for the design of intervention strategies to maintain functional β-cell mass as a treatment for T1D and T2D.

2. The Macrophage Activation Spectrum

Macrophages play an important role in maintaining tissue homeostasis, completing essential tissue-specific functions, and protecting the organism from infection. Due to the presence of scavenger receptors, they are able to perform housekeeping tasks such as removal of aged red blood cells, necrotic tissue, and toxic molecules, in the absence of special activation-associated stimuli. However, under the distress of infected or injured tissue, these homeostatic functions are increased by a variety of activating stimuli [8].

Tissue-resident macrophages were thought to continuously repopulate from circulating monocytes, which are ontologically derived from hematopoietic stem cells [9]. Recent studies have challenged this view. Although monocytes have the ability to differentiate into macrophages, subpopulations of resident macrophages in certain tissues (such as the pancreas) result from yolk-sac derived precursors during embryonic development [10,11]. This suggests that the pancreatic macrophage population is able to be maintained independently of circulating monocytes [12].

Macrophages are traditionally divided into two functional subgroups; the classically activated, inflammatory and cytotoxic M1-like macrophages and the alternatively activated M2-like macrophages that are anti-inflammatory and mediate tissue repair and remodeling [10]. It is now understood that these subsets better represent different points on a spectrum of macrophage activation states [13] and that other activation states may well be present [14]. Macrophages are able to respond to specific environmental signals to express various activation states along a dynamic range of phenotypes [15]. Nevertheless, for the sake of simplicity, we will use the subgroups M1-like and M2-like macrophage designations (Figure 1).

M1-like macrophages are characterized by their immunogenic properties. These properties protect against infection by propagating proinflammatory responses [16]. In the case of obesity-related pancreatic inflammation that is common in T2D, an abundance of free fatty acids (FFA), reactive oxygen species (ROS), islet amyloid polypeptide (IAPP), and proinflammatory cytokines increase M1-like macrophage polarization [17]. An M1-like macrophage can further propagate this response through the activation of the non-obese diabetic (NOD)-, LRR- and pyrin domain containing 3 (NLRP3) inflammasome. Once activated, the NLRP3 inflammasome secretes cytokines such as interleukin-1 beta (IL-1β) that induce M1-like polarization of other macrophages [18]. The proinflammatory M1-like state is also enhanced in the presence of interferon-gamma (IFN-γ), lipopolysaccharide (LPS), IL-1β, and tumor necrosis factor-alpha (TNFα). Additionally, M1-like macrophages are efficient producers of effector molecules such as ROS and reactive nitrogen species (NOS). They also produce inflammatory cytokines, such as IL-1β, IL-6, IL-12, and TNF-α, and chemokines such as chemokine (C-X-C motif) ligand 9 (CXCL9), CXCL10, and Chemokine (C-C motif) ligand 5 (CCL5). Additionally, M1-like macrophages express major histocompatibility complex (MHC) class II. These factors allow the M1-like macrophage to be recognized by and induce an immune response from T helper 1 (Th1) cells [19].

The M2-like macrophage primarily serves an anti-inflammatory and tissue regenerative purpose. M2-like macrophage polarization has a wider variety of activating stimuli. Anti-inflammatory signals such as IL-4, IL-13, macrophage colony-stimulating factor (CSF-1), and transforming growth factor beta (TGF-β) are associated with differentiation to the M2-like tissue regenerative and remodeling phenotype [20]. IL-4 and IL-13 were among the first to be discovered as inhibitors of the classical M1-like phenotype [16]. IL-21 is another important cytokine that drives M2-like polarization. Additionally, IL-33 is a cytokine of the IL-1 family that amplifies IL-13-induced polarization of macrophages to the M2-like phenotype. The M2-like phenotypes are characterized by the expression of chitinase 3 like 1

(CHI3L1), CCL24, CCL17, and arginase 1. M1-like macrophages metabolize arginine through NOS2 to produce nitric oxide (NO), M2-like macrophages use arginase 1 (ARG1) to metabolize arginine into polyamines that are used for proliferation and tissue remodeling [16]. M2-like macrophages express Th2 attracting chemokines CCL17 and CCL24, allowing the M2-like macrophages to work synergistically with Th2 cells to resolve inflammation and promote tissue remodeling, angiogenesis, and immunoregulation. Even within the M2-like "classification" there appear to be M2-like subtypes based on in vitro activation studies, expression profile, and function. M2a macrophages are activated by IL-4 and IL-13, express scavenger and phagocytic receptors, and secrete fibronectin, insulin-like growth factor (IGF), and TGFβ. M2b macrophages produce the proinflammatory IL-1β, IL-6, and TNF-α, and the anti-inflammatory IL-10. M2b cells are activated by Toll-like receptors and IL-1 receptor antagonists. M2c macrophages are activated through IL-10 and glucocorticoids. They remove apoptotic cells through expression of tyrosine-protein kinase MER (MerTK). Finally, M2d macrophages express IL-10 and vascular endothelial growth factor (VEGF). This wide variety of "M2" macrophage subtypes emphasizes the current understanding that M1-like and M2-like macrophages define extremes of a "macrophage activation spectrum" in terms of activation, gene expression, and function [21–24].

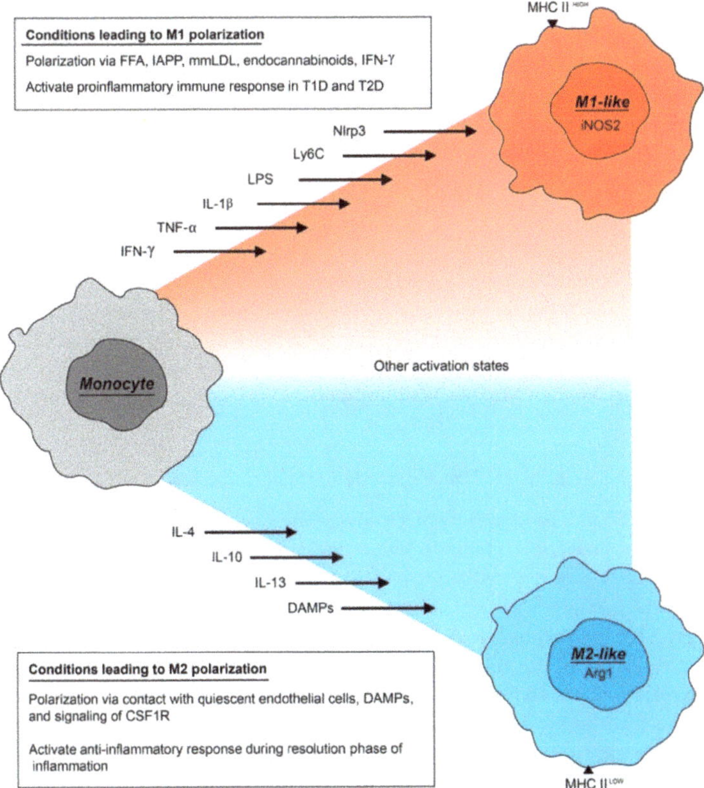

Figure 1. Polarization of monocytes to M1-like or M2-like macrophages. Macrophages can be polarized along an activation spectrum in response to different signals within their microenvironment. An undifferentiated macrophage will lean more M1-like under proinflammatory conditions, propagating a "kill" response. Conversely, a macrophage may lean more M2-like under anti-inflammatory conditions usually following M1 damage where a "heal" response is necessary.

3. Macrophages Can Impair β-Cell Function and Survival

Extensive findings demonstrate that M1-like macrophages have deleterious effects on β-cells by impairing glucose-stimulated insulin secretion (GSIS), inducing apoptosis, and causing β-cell dedifferentiation. NOD mice that spontaneously develop T1D, can have diabetes development impaired through macrophage depletion [25]. These findings have been substantiated in the BioBreeding Diabetes Prone rat (DP-BB) model [26]. Similarly, human tissue sections from T1D patients demonstrate macrophage infiltration of the islet [27,28]. The key proinflammatory cytokines observed in T1D are TNF-α, IL-1β, and IFN-γ, all of which are produced by macrophages [29]. Macrophages isolated from T1D patients have elevated proinflammatory gene expression. Interestingly, macrophages from long-term T1D patients have an impaired ability to undergo M2-like activation, suggesting potential changes to the macrophage population over time [30]. Macrophages exposed to oxidative stress produced elevated ROS levels that can push the macrophage toward an M1-like phenotype, which can be reverted to an M2-like phenotype with the use of ROS scavengers [31,32].

Islet inflammation is also a characteristic of T2D. Increased macrophage numbers have been observed in human T2D pancreata as well as rodent T2D models such as db/db mice, GK rats, and diet-induced obesity models [18,33–39]. Islets from T2D patients have elevated IL-6, IL-8, CXCL1, granulocyte colony-stimulating factor (GCSF), and macrophage inflammatory protein-1 alpha (MIP1α) levels [33]. Hyperglycemia and hyperlipidemia induce islet chemokine secretion which results in macrophage islet infiltration. These macrophages express M1-like markers, and depletion of M1-like macrophages suppresses β-cell lipotoxicity in vivo [9,36]. Various systemic changes associated with obesity can induce macrophage activation to an M1-like phenotype, and result in increased secretion of inflammatory cytokines such as IL-1β, IFN-γ, and TNF-α that directly signal to and affect β-cell function and survival [40]. Using clodronate-mediated macrophage suppression, db/db and KKAy mice have improved glucose tolerance and insulin secretion, demonstrating that macrophages in these T2D models impair β-cell function [36]. β-cell dedifferentiation is a hallmark of T2D and plays a major role in decreasing functional β-cell mass [41]. Db/db islets have impaired expression of β-cell identity genes, such as MAF BZIP Transcription Factor A (MafA), pancreatic and duodenal homeobox 1 (Pdx1), Glut2, and potassium inwardly rectifying channel subfamily J member 11 (Kcnj11). Interestingly, β-cell identity gene expression improves when macrophages are depleted from db/db islets [42]. These results demonstrate that M1-like macrophages have a deleterious effect on functional β-cell mass in T1D and T2D.

Macrophage Produced Secreted Factors That Negatively Modulate Functional β-Cell Mass

The primary cytokines produced by macrophages that negatively impact functional β-cell mass are IL-1β, IL-6, IFN-γ, and TNF-α (Table 1). These cytokines have negative effects on functional β-cell mass by impairing GSIS, inducing cell death, and increasing β-cell dedifferentiation. IL-1β is a member of the interleukin 1 (IL-1) family of cytokines. As such, IL-1β is involved early in the immune response to signal the production of other cytokines. Although the islet is capable of producing IL-1β under diabetogenic conditions, mouse and human islet studies have shown that macrophages secrete the overwhelming majority of IL-1β [43]. Additionally, β-cells express high levels of interleukin-1 receptor (IL-1R), and as such are highly sensitive to IL-1β [44]. It has been shown that when β-cells are subject to acute exposure of IL-1β, GSIS and overall β-cell survival improve. However, numerous studies have shown that chronic exposure to IL-1β common to diabetogenic or obese conditions causes impaired GSIS and increased β-cell de-differentiation and death [43,45,46]. Blocking IL-1β or inhibiting IL-1R results in improved outcomes in the restoration of β-cell function, mass, and the reversal of T2D phenotypes [9,47,48].

IL-1β signaling results in activation of nuclear factor kappa-light-chain-enhancer of activated B cells (NF-κB) and mitogen-activated protein kinase (MAPK) pathways [49], through which the β-cell releases chemokines such as CXCL1 and CXCL2 to recruit cells of the innate and adaptive immune system. In a study using neonatal NOD mice, those treated with a neutralizing anti-IL-1β monoclonal antibody (mAb) had decreased macrophage and neutrophil islet infiltration [50]. Similarly, mice treated with anti-IL-1β mAb demonstrated a significant decrease in islet produced CXCL1 and CXCL2, which recruits infiltrating immune cells through the CXCR2 receptor [51,52]. The use of IL-1R antagonist drugs has shown significant reductions in islet macrophage infiltration [53]. Finally, β-cells exposed to IL-1β increase expression and secretion of other proinflammatory signals monocyte chemoattractant protein-1 (MCP-1), IL-6, and TNF-α, thus inducing greater macrophage migration and islet inflammation [18]. The combined exposure of islets to IL-1β and IL-6 decreases GSIS, increases endoplasmic reticulum (ER) stress marker expression (such as Inducible nitric oxide synthase 2 (iNOS2), activating transcription factor 4 (ATF4), and CCAAT/enhancer binding protein (C/EBP) homologous protein (CHOP)), induces calcium handling deficiencies, and increases cell death [54]. These data demonstrate the IL-1β signaling at the β-cell results in the production of signals to enhance islet inflammation.

Culturing macrophages ex vivo with elevated glucose or fatty acids induces IL-1β release [47,55]. Min6 β-cells cultured with conditioned media from palmitate-treated macrophages demonstrated that preclearing the media with neutralizing antibodies blocked the effect of these cytokines to impair β-cell function [36]. Macrophages cultured with Min6 β-cells result in increased cytokine secretion, which impedes GSIS, and the addition of anti-IL-1β and anti-TNF-α antibodies improves GSIS [36]. β-cell IL-1β signaling through the MAPK and c-Jun N-terminal kinase (JNK) pathway results in downregulation of the phosphatidylinositol 3-kinase-protein kinase B signaling pathway (P13K-AKT) signaling cascade, and ultimately decreases PDX1-mediated gene expression [45,56]. This results in a decrease in insulin mRNA levels and impaired GSIS [57–63].

IL-1β induces pro-apoptotic and necrotic pathways in the β-cell through the extracellular signal-regulated kinase (ERK) signaling pathways [64]. ER stress that is characteristic of T2D β-cells potentiates the IL-1β signaling pathway and leaves the β-cell more susceptible to IL-1β-mediated cell death [54]. TNF-α and IFN-γ work synergistically with IL-1β to induce β-cell apoptosis via the intrinsic and extrinsic apoptotic pathways. IL-1β and IFN-γ induce the intrinsic apoptotic pathway through the activation of the NF-κB-mediated gene network. NF-κB activation subsequently leads to NO and cytokine production, depletion of ER calcium stores, and induction of ER stress. Rat islets cultured with IL-1β and IFN-γ revealed expression changes to genes associated with inflammation, cell death, antigen presentation, and cytokines/chemokine production [65]. ER stress results in mitochondrial damage, cytochrome c release, and mitochondrial death signals that activate caspase 9 and caspase 3 resulting in activation of the intrinsic apoptosis pathway [66,67]. Human islets cultured with TNF-α, IFN-γ, and IL-1β with or without NO induced ER stress as indicated by increased expression of CHOP, activating transcription factor 3 (ATF3), binding immunoglobulin protein (BIP), and X-box binding protein-1 (XBP1). IL-1β can also induce apoptosis via the extrinsic pathway by up-regulating Fas receptor expression [68].

IL-1β can induce β-cell dedifferentiation. Human and rodent islets cultured with IL-1β, IL-6, and TNF-α present with β-cell dedifferentiation [69]. Furthermore, IL-1β induces downregulation of forkhead box protein O1 (Foxo1), which is essential for maintaining β-cell differentiation. Similar results are observed with EndoC-βH1 and human islets, where culture with the same mixture of cytokines induces upregulation of progenitor genes such as SRY-Box Transcription Factor 9 (Sox9), and downregulation of mature β-cell genes [70]. Interestingly, culturing β-cells with non-cytotoxic IL-1β levels impaired insulin secretion, reduced β-cell proliferation, and decreased expression of β-cell identity genes such as MafA and Urocortin-3 (Ucn3) [71]. These changes, however, were reversible as IL-1β removal restored β-cell identity gene expression. While these data strongly suggest that IL-1β plays a critical role in inducing β-cell dedifferentiation, pancreatic IL-1R deletion (the receptor by which IL-1β signal transduction occurs) also results in impaired β-cell function and increased expression of the dedifferentiation marker Aldh1a3 [45]. More studies will need to be completed to clearly understand the role of IL-1β on β-cell dedifferentiation.

One of the key differences between M1-like and M2-like macrophages is arginine metabolism. M1-like macrophages use arginine to produce NO by way of iNOS2. Macrophage production of NO and ROS have direct effects on functional β-cell mass due to low β-cell expression of radical scavenging pathways. NF-κB regulates the expression of inducible nitric oxide synthase (iNOS) in β-cells, with many of the gene expression changes associated with cytokine exposure being secondary to iNOS-mediated NO formation. Increased macrophage produced NO and ROS leads to DNA damage and activation of poly (ADP-ribose) polymerase (PARP) to facilitate DNA repair. PARP activation depletes the β-cells NAD pool, which can lead to β-cell necrosis [72]. IL-1β stimulates β-cell iNOS expression, which leads to elevated internal NO levels. Increased cellular NO impairs electron transfer, decreases mitochondrial ATP production, and induces the expression of proinflammatory genes in the β-cell [73].

IL-6 signals through the IL-6 receptor system. This results in signal transducer and activator of transcription 3 (STAT3) activation and MAPK activation and increased transcription of downstream target genes [74,75]. IL-6 is sufficient to impair GSIS and decrease islet insulin content [76,77]. Conversely, chemically inhibiting IL-6 signaling at the IL-6 receptor improves insulin content and GSIS [78]. The observed decrease in insulin content due to IL-6 is due to transcriptional changes, as IL-6 decreases Ins1, Ins2, and PDX1 mRNA levels [79]. IL-6 signaling may also impair mitochondrial function through inducing mitochondrial fission and potentially increased mitophagy [80].

IFN-γ is a pro-inflammatory cytokine responsible for β-cell destruction. IFN-γ signals to the β-cell through the interferon-gamma receptor (IFNGR) complex. This receptor complex activates the JAK/STAT and the NF-κB signaling cascades [49,81]. This leads to activation of the transcription factor interferon regulatory factor 1 (IRF-1) and upregulation of caspase-1, caspase-3, caspase-9 [82], and other proapoptotic gene expression [82]. Loss of STAT1 signaling impairs IFN-γ-mediated cell death [83]. Furthermore, loss of STAT1 impairs IP-10 and iNOS gene expression [83]. IFN-γ also impairs β-cells insulin content and GSIS [84,85]. Finally, IFN-γ potentiates IL-1β-mediated iNOS expression and NO production in the β-cell [86–88].

TNF-α was originally thought to be produced by T-cells, however, data demonstrate that TNF-α is primarily produced by macrophages and dendritic cells in the islet [89,90]. TNF-α signals through the TNFR and activates the NF-κB and MAPK signaling cascades [49,91]. Once a threshold of activation through NF-κB is activated, proapoptotic and inflammatory pathways are activated [92]. Inhibition of TNF-α, NF-κB, and JNK has been shown to improve β-cell survival and function [93]. These signaling cascades can lead to β-cell dysfunction and death [93]. TNF-α, IL-1β, and IFN-γ all induce ROS production in β-cells through activation of nicotinamide adenine dinucleotide phosphate (NADPH) oxidases [94]. They also upregulate iNOS expression, resulting in increased NO production [95,96]. Increased ROS and NO production results in mitochondrial damage and activation of the intrinsic apoptotic pathway [97]. Induction of iNOS through the TNF-α signaling cascade functions through the IRF transcription factor [98,99]. Human islets exposed to TNF-α, IL-1-β, and IFN-γ have increased expression of various cytokines that enhance immune cell migration, including CXCL1, CXCL8, CCL20, CCL2, and CXCL10 through the NF-κB and STAT1 signaling cascade [100]. TNF-α also induces β-cell Ca^{2+} influx, which can negatively affect insulin secretion and β-cell survival [101]. Finally, using a TNF-α antagonist partially restored β-cell identity gene expression that are lost during dedifferentiation [69]. These data demonstrate the negative effect that these various M1-like macrophages produced secreted factors have on the pancreatic β-cell.

Table 1. Macrophages secreted factors that have a negative effect on functional β-cell mass.

Effectors	Target
IL-1β	Initiates β-cell apoptosis through ERK signaling pathways [18,64] Decreases insulin mRNA levels [57–63] Impairs GSIS [57–63] Increases IL-6 release in β-cell [18] Transcriptional changes of 3068 genes associated with inflammation, cell death, antigen presentation, and cytokines/chemokines [65] Contributes to increased ER stress [68] Increases Fas expression [77]
IL-6	Impairs GSIS [76,77] Decreases Ins1, Ins2, and PDX1 mRNA levels in the islet [79]
IFN-γ	Participates with IL-1β to activate NF-κB genes, leading to NO and cytokine production leading to ER stress [65] Participates with IL-1β to cause transcriptional changes to 3068 genes associated with inflammation, cell death, antigen presentation, and cytokines/chemokines [65] Impairs β-cell insulin secretion [84,85]
TNF-α	Contributes to increased ER stress [68] Participates in NF-κB pathway activation [92] Activates proapoptotic and proinflammatory pathways through NF-κB [49,91,93] Increases iNOS and NADPH oxidase activity, leading to increased ROS production and mitochondrial damage [94–97] Induces intrinsic apoptosis [97] Increases the expression of cytokines CXCL1, CXCL8, CCL20, CCL2, and CXCL10, which promote immune cell infiltration of the islet [100] Induces Ca^{2+} influx in β-cells, impairing insulin secretion [101]

4. M2-Like Macrophage Can Enhance the Development, Maintenance, and Function of β-Cells

There is an intimate relationship between macrophages and the development of pancreatic islets, and specifically β-cell proliferation and survival. Mature F4/80+ macrophages are found in the pancreatic bud by E14.5 [102]. The presence of these macrophages is directly linked to the expansion of β-cell number during development. Pancreas explants cultured with exogenous macrophage colony-stimulating factor (M-CSF), which is sufficient to induce macrophage proliferation, results in increased macrophage and β-cell number more than four-fold over the explants cultured without M-CSF, suggesting a connection between macrophage signaling and β-cell proliferation [102]. The presence and increased migration of macrophages to the developing pancreas is necessary for the delamination of endocrine cells from the pancreatic ducts and their ultimate migration to nascent islets [103]. Similarly, macrophages are observed in the human pancreas as early as 6 weeks of development, with elevated levels of the cytokine CSF-1 also being detected [104]. As CSF-1 is essential for macrophage differentiation, these data suggest that macrophages are needed for nascent β-cell expansion. CSF-1-deficient osteopetrotic CSF-1 op/op mice, which have impaired production of myeloid-derived macrophages and dendritic cells, also have decreased β-cell mass and impaired β-cell proliferation when compared to CSF-1 op/+ littermates [102,103]. These data demonstrate that macrophages are necessary for β-cell proliferation during embryonic development. Furthermore, given the function of these macrophages, these data suggest a M2-like phenotype.

While the data for macrophage-mediated β-cell proliferation during development are strong, there are equally compelling data regarding the effects of the macrophage on β-cell survival and proliferation during adulthood or during disease states. Mice fed a high-fat diet initially demonstrate increased β-cell proliferation. This proliferation correlates with increased intra-islet macrophage accumulation, where these macrophages express markers indicative of an M2-like activation state [105,106]. In fact, hyperplastic islets observed in diet-induced obesity have greater macrophage concentrations, suggesting that macrophages may be needed to open the extracellular matrix and create an expansion niche for the growing islet [33]. These studies are supported by other models that demonstrate the effect of macrophages on the β-cell. Macrophages with an M1-like expression profile shifted to an M2-like profile in response to diphtheria toxin-mediated β-cell damage and in correlation with the subsequent β-cell expansion [107,108]. Clodronate-mediated macrophage ablation attenuates β-cell regeneration, suggesting that macrophages are needed for β-cell expansion in this model. Similarly, pancreatic ductal ligation-mediated β-cell regeneration is dependent on M2-like macrophages [109–112]. Using a chronic pancreatitis model of β-cell loss, it was shown that M2-like activated macrophages were essential for β-cell proliferation and that transplantation with CSF-1R -/- bone marrow (which has impaired macrophage production) results in lost β-cell mass [113]. Changes in vascularization were also observed, and given the ability of macrophages to assist in vascular remodeling, some of the β-cell maintenance and proliferation may be vascular mediated [20]. β-cell regeneration after VEGF-A-mediated β-cell loss was shown to be dependent on M2-like macrophages [114,115]. Finally, recent observations demonstrate that human pancreatic donors demonstrated a strong correlation between the presence of M2-like macrophages and increased islet vascularization and β-cell proliferation marker expression [116]. These data demonstrate a direct connection between M2-like macrophages and the maintenance and expansion of β-cell mass in various models of β-cell damage and regeneration.

Macrophage Produced Secreted Factors That Positively Modulate Functional β-Cell Mass

M2-like macrophages clearly play a protective role in β-cells in terms of maintaining and expanding functional β-cell mass. This protective role is mediated by various factors secreted from the macrophage that directly affect the β-cell (Table 2). Using a streptozotocin (STZ) model of β-cell destruction, recruitment of M2-like macrophage to islets was observed. It was shown that these M2-like macrophages produce and release wingless-type MMTV integration site family, member 3A (Wnt3a), which induces the Wnt/β-catenin pathway in the β-cell, which resulted in increased β-cell proliferation [117]. Similarly, using a diphtheria toxin-induced model of β-cell injury, M2-like macrophages were shown to induce β-cell survival and proliferation through macrophage produced Wnt and activation of the β-cell Wnt/β-catenin signaling pathway [107,118].

Resident macrophages can be moved to an M2-like phenotype through IL-33 signaling. In response to islet secreted IL-33 and the change of the macrophage phenotype to an M2-like state, changes in secreted factors are observed. This signaling cascade, also facilitated by IL-13 and CSF-2, induces expression of aldehyde dehydrogenase 1 family, member A2 (Aldh1a2), which produces retinoic acid. Retinoic acid released from resident macrophages (and dendritic cells) results in the induction of retinoic acid receptor beta (RARβ) expression in the β-cell and increased insulin [119].

Using a partial pancreatic ductal ligation demonstrates additional macrophage secreted factors that affect functional β-cell mass. M2-like macrophages were shown to migrate in response to ductal ligation and secrete TGFβ1 and EGF. TGFβ1 induced β-cell upregulation of mothers against decapentaplegic homolog 7 (SMAD7) and SMAD2, where SMAD7 is sufficient to induce proliferation and SMAD2 acts as an inhibitor of the SMAD7 proliferation cascade. In addition to secreting TGFβ1, the M2-like macrophages also secreted EGF. The EGF signaling cascade resulted in inhibition of SMAD2 nuclear localization. This allowed β-cells to undergo proliferation through upregulation of the cell cycle activators cyclin D1/D2 and nuclear exclusion of the cell cycle inhibitor p27^{CDKN1B} [111]. TGFβ has also been shown to be essential for proper pancreas development [120,121]. Furthermore, knockout of TGFβ receptor I and II in β-cells inhibits M2-like macrophages-mediated β-cell proliferation [111,121].

EGF is not the only growth factor secreted from macrophages that have been shown to have a positive effect on β-cells. Macrophages from obese mice have been shown to increase β-cell proliferation. This occurs due to PDGF release from the macrophage and signaling to the β-cell through platelet-derived growth factor receptor (PDGFR) [122]. Hepatocyte growth factor (HGF) has been shown to push macrophages to an M2-like phenotype [123], and M2-like macrophages are capable of secreting HGF [10]. HGF has been shown to increase β-cell proliferation and increase insulin secretion, as well as induce β-cell regeneration in a partial pancreatectomy model [124,125]. IGF-1 produced from M-2 like macrophages has been shown to play a critical role in maintaining functional β-cell mass by maintaining glucose-stimulated insulin secretion [126].

In addition to producing growth factors, M2-like macrophages release IL-10, which has direct effects on maintaining functional β-cell mass. β-cells exposed to IL-10 express elevated levels of anti-apoptotic genes [127]. IL-10 can induce β-cell iNOS protein levels, thus potentially decreasing NO levels [128]. Finally, IL-10 can enhance insulin secretion, as measured by c-peptide levels [129]. A recent publication showed that human islets have macrophages in the perivascular region that are the primary source of islet IL-10 levels, and that loss of the ability to produce IL-10 in the obese and diabetic state leads to β-cell loss [130].

Finally, islet expansion is dependent on the remodeling of the islet vasculature and the extracellular matrix (ECM), as well as the release of growth factors bound in the ECM. M2-like macrophages produce and secrete matrix metalloproteinases (MMP) that are critical for this process. Macrophages from human islets are the primary source of MMP9 [130]. Macrophages that migrate to the islets in a chronic pancreatitis model of diabetes express various M2-like markers, including MMP9, which is essential for islet vascularization and β-cell expansion [113]. Together, these data demonstrate the various factors secreted from the M2-like macrophages that have direct effects on maintaining functional β-cell mass.

Table 2. Macrophages secreted factors that have a positive effect on functional β-cell mass.

Effectors	Target
WNT3A	Increases β-cell proliferation and survival via Wnt/β-catenin pathway [107,117,118]
RETINOIC ACID	Increased expression of RARβ and increased insulin production and secretion [119]
TGFβ1	Induces upregulation of SMAD7, which is responsible for increased β-cell proliferation [111] Induces upregulation of SMAD2, which is a SMAD7 inhibitor [111]
EGF	Inhibits SMAD2 nuclear localization, working in conjunction with TGFβ1 to induce β-cell proliferation [111]
PDGF	Induces β-cell proliferation [122]
IGF-1	Promotes β-cell survival by maintaining GSIS [126]
IL-10	Induces upregulation of anti-apoptotic genes, promoting greater β-cell survival [127] Increases iNOS levels, decreasing NO levels in β-cells [128] Increases insulin secretion [129]
MMP9	Promotes islet vascularization and β-cell expansion [113]

5. Use of Macrophages to Improve Functional β-Cell Mass as a Treatment for Diabetes

The previous studies demonstrate the potential for M2-like macrophages to be harnessed to ameliorate functional β-cell mass as a treatment for T1D and T2D. Various studies have begun to demonstrate this feasibility. While macrophages from long-term T1D patients have impaired ability to undergo M2-like activation [30], NOD mice treated with the BET antagonist I-BET151 demonstrated significant M2-like activation and β-cell regeneration [131]. Similarly, NOD mice that received adoptive transfer of M2-like macrophages resulted in greater than 80% protection against T1D phenotypes for over 3 months compared to NOD mice that did not receive adoptive macrophage transfer or mice that received adoptive transfer of non-M2-like macrophages [132]. In this model, it was shown that the adoptively transferred M2-like macrophages preferentially migrated to the pancreas, where a greater number of islets were maintained after M2-like macrophage adoptive transfer. Using an STZ model of β-cell injury, it was shown that adoptive transfer of M2-like macrophages was sufficient to protect functional β-cell mass and diabetes-induced kidney damage. Mice that received M2-like macrophage adoptive transfer had greater islet area, as well as improved HbA1c and non-fasting blood glucose levels [133]. Finally, STZ mice transplanted with mesenchymal stem cells resulted in improved fasting blood glucose, β-cell area, and β-cell proliferation. This corresponded with increased infiltration of M2-like macrophages [117]. It was shown that the mesenchymal stem cells were inducing endogenous macrophages to undergo M2-like activation. The M2-like macrophages induced β-cell proliferation through induction of the Wnt3a/β-catenin signaling pathway. Although these preliminary rodent studies demonstrate the ability of M2-like macrophages to block or reverse diabetes progression and loss of functional β-cell mass, future studies in humans will be required.

6. Concluding Remarks

The purpose of this review was to highlight the detrimental and beneficial effects of macrophages, based on their activation state, on functional β-cell mass. From the studies reviewed here, it is clear that macrophages can have opposing effects on the β-cell. The various signals from the islet, from other autoimmune cells, and from systemic inflammation present in either T1D or T2D have the potential to induce M1-like activation of the macrophage and macrophage-mediated decrease in functional β-cell mass. Alternatively, signals from the β-cell under normal physiology or in response to certain forms of β-cell damage can result in M2-like macrophage activation and secretion of factors that promote β-cell proliferation, survival, and insulin secretion.

One potential caveat of these findings is the applicability to human β-cells. Given the physiological differences between rodent and human islets, it will be essential to validate the beneficial effects of M2-like macrophages that have been observed in rodent islets in human islets. Rodent islets or mice were the primary models used in the majority of studies cited in this review (Table 3). Fortunately, there are more and more studies looking at the effect of macrophages on human islets, or exploring the role of macrophages in vivo on human functional β-cell mass. For the greater application of these findings, these studies will need to be expanded.

Table 3. References that studied the effect of M1-like or M2-like macrophages on rodent islets, human islets, or both.

Rodent or Human Islets	References
Rodent	[18,25,26,31–34,37,38,40,42,45–47,50–52,55,56,58,62,64,67,71,73,79,86,89,94,98,99, 102,103,106–108,111,113,115,116,118,119,121,123,124,127,132,133]
Human	[27,28,30,36,39,43,48,61,63,69,77,78,100,104,117,130,131]
Both	[5,17,20,59,66,76,93,97,120]

A clearer understanding of the processes of macrophage activation and the macrophage produced factors that directly affect the β-cell could be utilized as a potential therapeutic to maintain and expand functional β-cell mass as a treatment for both T1D and T2D (Figure 2). Furthermore, understanding the differences between resident and recruited macrophages and their ability to maintain or damage β-cells may be leveraged to improve patient care. In addition to having direct application to T1D, these findings may be applicable to other autoimmune disorders. Similarly, understanding the macrophage-mediated changes observed at the pancreatic islet in T2D may be beneficial for other obesity-related pathologies in other organs.

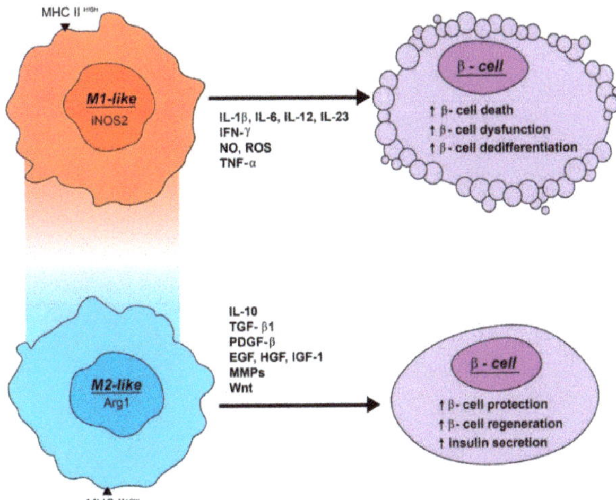

Figure 2. How the M1-like and M2-like extremes of the macrophage activation spectrum affect β-cell function, survival, and proliferation through secreted factors. M1-like macrophages are inflammatory in nature, producing and secreting inflammatory cytokines, NO, and ROS. M2-like macrophages play more of a protective role by inducing β-cell survival and proliferation. M2-like macrophages secrete anti-inflammatory cytokines, signaling peptides, and enzymes involved in tissue remodeling, inducing β-cell proliferation and enhancing insulin secretion.

Author Contributions: D.M.J., K.V.H., A.T.M., and J.S.T. each contributed to the investigation and writing of this manuscript. All authors have read and agreed to the published version of the manuscript.

Funding: This research was funded by The Beatson Foundation, grant number 2019-003 (J.S.T.).

Acknowledgments: The authors thank the members of the Tessem lab at BYU for critical discussion regarding these topics.

Conflicts of Interest: The authors declare no conflict of interest.

Abbreviations

Aldh1a2	Aldehyde dehydrogenase 1 family, member A2
ARG1	Arginase 1
ATF3	Activating transcription factor 3
ATF4	Activating transcription factor 4
ATP	Adenosine triphosphate
BET	Bromodomain and extraterminal domain family
BIP	Binding immunoglobulin protein
CCL17	Chemokine (C-C motif) ligand 17
CCL20	Chemokine (C-C motif) ligand 20
CCL24	Chemokine (C-C motif) ligand 24
CCL5	Chemokine (C-C motif) ligand 5
CHI3L1	Chitinase 3 Like 1
CHOP	CCAAT/enhancer binding protein (C/EBP) homologous protein
CSF-1	Macrophage colony-stimulating factor
CSF-1R	Colony-stimulating 1 receptor
CSF-2	Granulocyte-macrophage colony-stimulating factor
CXCL1	Chemokine (C-X-C motif) ligand 1
CXCL10	Chemokine (C-X-C motif) ligand 10 (see also IP-10)
CXCL2	Chemokine (C-X-C motif) ligand 2
CXCL8	Chemokine (C-X-C motif) ligand 8

CXCL9   Chemokine (C-X-C motif) ligand 9
CXCR2   Chemokine (C-X-C motif) receptor 2
Db/db   Diabetic mouse model
DNA   Deoxyribonucleic acid
DP-BB   BioBreeding Diabetes Prone rat
ECM   Extracellular matrix
EGF   Endothelial growth factor
ER   Endoplasmic reticulum
ERK   Extracellular signal-regulated kinase
Fas   Apoptosis antigen 1
Foxo1   Forkhead box protein O1
GCSF   Granulocyte colony-stimulating factor
GK   Goto-Kakizaki rat
GSIS   Glucose-stimulated insulin secretion
HGF   Hepatocyte growth factor
I-BET151   BET bromodomain inhibitor
IFN-γ Interferon-gamma
IFNGR   Interferon gamma receptor
IGF   Insulin-like growth factor
IGF-1   Insulin-like growth factor 1
IL-1   Interleukin-1
IL-12   Interleukin-12
IL-13   Interleukin-13
IL-18   Interleukin-18
IL-1β Interleukin-1 beta
IL-21   Interleukin-21
IL-33   Interleukin-33
IL-4   Interleukin-4
IL-6   Interleukin-6
IL-8   Interleukin-8
iNOS   Inducible nitric oxide synthase
iNOS2   Inducible nitric oxide synthase 2
IP-10   Chemokine (C-X-C motif) ligand 10 (see also CXCL10)
IRF-1   Interferon regulatory factor 1
Isl-1   Insulin gene enhancer protein 1
JAK/STAT   Janus kinase/signal transducer and activator of transcription
JNK   c-Jun N-terminal kinase
Kcnj11   Potassium Inwardly Rectifying Channel Subfamily J Member 11
LPS   Lipopolysaccharide
mAb   Monoclonal antibody
MafA   MAF BZIP Transcription Factor A
MAPK   Mitogen-activated protein kinase
MCP-1   Monocyte chemoattractant protein-1
M-CSF   Macrophage colony-stimulating factor
MerTK   Tyrosine-protein kinase MER
Min6   Mouse insulinoma cell line
MIP1α Macrophage inflammatory protein-1 alpha
MMP   Metalloproteinase
MMP9   Metalloproteinase 9
mRNA   Messenger ribonucleic acid
NAD   Nicotinamide adenine dinucleotide
NADPH   Nicotinamide adenine dinucleotide phosphate
NF-κB   Nuclear factor kappa-light-chain-enhancer of activated B cells
NLRP3   NOD-, LRR- and pyrin domain containing 3
NO   Nitric oxide
NOD   Non-obese diabetic

NOS	Nitrogen species
NOS2	Nitric oxide synthase
p27^{CDKN1B}	Cyclin-dependent kinase inhibitor 1B
PARP	Poly (ADP-ribose) polymerase
PDGF	Platelet-derived growth factor
PDGFR	Platelet-derived growth factor receptor
PDX1	Pancreatic and duodenal homeobox 1
PI3K-AKT	Phosphatidylinositol 3-kinase-protein kinase B signaling pathway
RARβ	Retinoic acid receptor beta
ROS	Reactive oxygen species
SMAD2	Mothers against decapentaplegic homolog 2
SMAD7	Mothers against decapentaplegic homolog 7
Sox9	SRY-Box Transcription Factor 9
STAT1	Signal transducer and activator of transcription 1
STAT3	Signal transducer and activator of transcription 3
STZ	Streptozotocin
T1D	Type 1 diabetes
T2D	Type 2 diabetes
TGFβ	Transforming growth factor alpha
TGF-β1	Transforming growth factor beta 1
Th1	T-helper type 1 cell
Th2	T-helper type 2 cell
TNF-a	Tumor necrosis factor alpha
TNFR	Tumor necrosis factor receptor
Ucn3	Urocortin-3
VEGF	Vascular endothelial growth factor
VEGF-A	Vascular endothelial growth factor A
Wnt3a	Wingless-Type MMTV Integration Site Family, Member 3A
XBP1	X-box binding protein-1
YM1	Synonym for chitinase-like 3
ZDF	Zucker diabetic fatty rat

Note: References [5,30,33,37–39,43,47,58,69,76,102–104,107,113,115–117,122,130–133] indicates key finding in the field, relative to this topic.

References

1. Saeedi, P.; Petersohn, I.; Salpea, P.; Malanda, B.; Karuranga, S.; Unwin, N.; Colagiuri, S.; Guariguata, L.; Motala, A.A.; Ogurtsova, K.; et al. Global and regional diabetes prevalence estimates for 2019 and projections for 2030 and 2045: Results from the International Diabetes Federation Diabetes Atlas, 9th edition. *Diabetes Res. Clin. Pr.* **2019**, *157*. [CrossRef] [PubMed]

2. Mallone, R.; Eizirik, D.L. Presumption of innocence for beta cells: Why are they vulnerable autoimmune targets in type 1 diabetes? *Diabetologia* **2020**, *63*, 1999–2006. [CrossRef] [PubMed]

3. Efrat, S. Beta-Cell Dedifferentiation in Type 2 Diabetes: Concise Review. *Stem Cells* **2019**, *37*, 1267–1272. [CrossRef] [PubMed]

4. Pugliese, A. Insulitis in the pathogenesis of type 1 diabetes. *Pediatr. Diabetes* **2016**, *17*, 31–36. [CrossRef]

5. Ying, W.; Fu, W.X.; Lee, Y.S.; Olefsky, J.M. The role of macrophages in obesity-associated islet inflammation and beta-cell abnormalities. *Nat. Rev. Endocrinol.* **2020**, *16*, 81–90. [CrossRef]

6. Saunders, D.; Powers, A.C. Replicative capacity of beta-cells and type 1 diabetes. *J. Autoimmun.* **2016**, *71*, 59–68. [CrossRef]

7. Ley, K. M1 Means Kill; M2 Means Heal. *J. Immunol.* **2017**, *199*, 2191–2193. [CrossRef]

8. Poltavets, A.S.; Vishnyakova, P.A.; Elchaninov, A.V.; Sukhikh, G.T.; Fatkhudinov, T.K. Macrophage Modification Strategies for Efficient Cell Therapy. *Cells* **2020**, *9*, 1535. [CrossRef]

9. Eguchi, K.; Nagai, R. Islet inflammation in type 2 diabetes and physiology. *J. Clin. Investig.* **2017**, *127*, 14–23. [CrossRef]

10. Van Gassen, N.; Staels, W.; Van Overmeire, E.; De Groef, S.; Sojoodi, M.; Heremans, Y.; Leuckx, G.; Van de Casteele, M.; Van Ginderachter, J.A.; Heimberg, H.; et al. Concise Review: Macrophages: Versatile Gatekeepers During Pancreatic beta-Cell Development, Injury, and Regeneration. *Stem Cells Transl. Med.* **2015**, *4*, 555–563. [CrossRef]

11. Schulz, C.; Gomez Perdiguero, E.; Chorro, L.; Szabo-Rogers, H.; Cagnard, N.; Kierdorf, K.; Prinz, M.; Wu, B.; Jacobsen, S.E.; Pollard, J.W.; et al. A lineage of myeloid cells independent of Myb and hematopoietic stem cells. *Science* **2012**, *336*, 86–90. [CrossRef] [PubMed]

12. Ginhoux, F.; Guilliams, M. Tissue-Resident Macrophage Ontogeny and Homeostasis. *Immunity* **2016**, *44*, 439–449. [CrossRef] [PubMed]

13. Jezek, P.; Jaburek, M.; Holendova, B.; Plecita-Hlavata, L. Fatty Acid-Stimulated Insulin Secretion vs. Lipotoxicity. *Molecules* **2018**, *23*, 1483. [CrossRef] [PubMed]

14. Hume, D.A. The Many Alternative Faces of Macrophage Activation. *Front. Immunol.* **2015**, *6*, 370. [CrossRef] [PubMed]

15. Orecchioni, M.; Ghosheh, Y.; Pramod, A.B.; Ley, K. Macrophage Polarization: Different Gene Signatures in M1(LPS+) vs. Classically and M2(LPS-) vs. Alternatively Activated Macrophages. *Front. Immunol.* **2019**, *10*, 1084. [CrossRef] [PubMed]

16. Mantovani, A.; Biswas, S.K.; Galdiero, M.R.; Sica, A.; Locati, M. Macrophage plasticity and polarization in tissue repair and remodelling. *J. Pathol.* **2013**, *229*, 176–185. [CrossRef] [PubMed]

17. Donath, M.Y.; Dalmas, E.; Sauter, N.S.; Boni-Schnetzler, M. Inflammation in obesity and diabetes: Islet dysfunction and therapeutic opportunity. *Cell Metab.* **2013**, *17*, 860–872. [CrossRef]

18. Jourdan, T.; Godlewski, G.; Cinar, R.; Bertola, A.; Szanda, G.; Liu, J.; Tam, J.; Han, T.; Mukhopadhyay, B.; Skarulis, M.C.; et al. Activation of the Nlrp3 inflammasome in infiltrating macrophages by endocannabinoids mediates beta cell loss in type 2 diabetes. *Nat. Med.* **2013**, *19*, 1132–1140. [CrossRef]

19. Liu, G.; Yang, H. Modulation of macrophage activation and programming in immunity. *J. Cell Physiol.* **2013**, *228*, 502–512. [CrossRef]

20. Aamodt, K.I.; Powers, A.C. Signals in the pancreatic islet microenvironment influence beta-cell proliferation. *Diabetes Obes. Metab.* **2017**, *19*, 124–136. [CrossRef]

21. Zizzo, G.; Hilliard, B.A.; Monestier, M.; Cohen, P.L. Efficient clearance of early apoptotic cells by human macrophages requires M2c polarization and MerTK induction. *J. Immunol.* **2012**, *189*, 3508–3520. [CrossRef] [PubMed]

22. Mosser, D.M.; Edwards, J.P. Exploring the full spectrum of macrophage activation. *Nat. Rev. Immunol.* **2008**, *8*, 958–969. [CrossRef] [PubMed]

23. Ferrante, C.J.; Leibovich, S.J. Regulation of Macrophage Polarization and Wound Healing. *Adv. Wound Care* **2012**, *1*, 10–16. [CrossRef] [PubMed]

24. Grinberg, S.; Hasko, G.; Wu, D.; Leibovich, S.J. Suppression of PLCbeta2 by endotoxin plays a role in the adenosine A(2A) receptor-mediated switch of macrophages from an inflammatory to an angiogenic phenotype. *Am. J. Pathol.* **2009**, *175*, 2439–2453. [CrossRef] [PubMed]

25. Burg, A.R.; Tse, H.M. Redox-Sensitive Innate Immune Pathways During Macrophage Activation in Type 1 Diabetes. *Antioxid. Redox Signal.* **2018**, *29*, 1373–1398. [CrossRef] [PubMed]

26. Oschilewski, U.; Kiesel, U.; Kolb, H. Administration of silica prevents diabetes in BB-rats. *Diabetes* **1985**, *34*, 197–199. [CrossRef]

27. Hanninen, A.; Jalkanen, S.; Salmi, M.; Toikkanen, S.; Nikolakaros, G.; Simell, O. Macrophages, T cell receptor usage, and endothelial cell activation in the pancreas at the onset of insulin-dependent diabetes mellitus. *J. Clin. Investig.* **1992**, *90*, 1901–1910. [CrossRef]

28. Gaglia, J.L.; Guimaraes, A.R.; Harisinghani, M.; Turvey, S.E.; Jackson, R.; Benoist, C.; Mathis, D.; Weissleder, R. Noninvasive imaging of pancreatic islet inflammation in type 1A diabetes patients. *J. Clin. Investig.* **2011**, *121*, 442–445. [CrossRef]

29. Nunemaker, C.S. Considerations for Defining Cytokine Dose, Duration, and Milieu That Are Appropriate for Modeling Chronic Low-Grade Inflammation in Type 2 Diabetes. *J. Diabetes Res.* **2016**, *2016*, 2846570. [CrossRef]

30. Juhas, U.; Ryba-Stanislawowska, M.; Brandt-Varma, A.; Mysliwiec, M.; Mysliwska, J. Monocytes of newly diagnosed juvenile DM1 patients are prone to differentiate into regulatory IL-10(+) M2 macrophages. *Immunol. Res.* **2019**, *67*, 58–69. [CrossRef]

31. El Hadri, K.; Mahmood, D.F.; Couchie, D.; Jguirim-Souissi, I.; Genze, F.; Diderot, V.; Syrovets, T.; Lunov, O.; Simmet, T.; Rouis, M. Thioredoxin-1 promotes anti-inflammatory macrophages of the M2 phenotype and antagonizes atherosclerosis. *Arterioscler. Thromb. Vasc. Biol.* **2012**, *32*, 1445–1452. [CrossRef] [PubMed]

32. Padgett, L.E.; Burg, A.R.; Lei, W.; Tse, H.M. Loss of NADPH oxidase-derived superoxide skews macrophage phenotypes to delay type 1 diabetes. *Diabetes* **2015**, *64*, 937–946. [CrossRef] [PubMed]

33. Ehses, J.A.; Perren, A.; Eppler, E.; Ribaux, P.; Pospisilik, J.A.; Maor-Cahn, R.; Gueripel, X.; Ellingsgaard, H.; Schneider, M.K.; Biollaz, G.; et al. Increased number of islet-associated macrophages in type 2 diabetes. *Diabetes* **2007**, *56*, 2356–2370. [CrossRef]

34. Masters, S.L.; Dunne, A.; Subramanian, S.L.; Hull, R.L.; Tannahill, G.M.; Sharp, F.A.; Becker, C.; Franchi, L.; Yoshihara, E.; Chen, Z.; et al. Activation of the NLRP3 inflammasome by islet amyloid polypeptide provides a mechanism for enhanced IL-1beta in type 2 diabetes. *Nat. Immunol.* **2010**, *11*, 897–904. [CrossRef]

35. Masters, S.L.; Mielke, L.A.; Cornish, A.L.; Sutton, C.E.; O'Donnell, J.; Cengia, L.H.; Roberts, A.W.; Wicks, I.P.; Mills, K.H.; Croker, B.A. Regulation of interleukin-1beta by interferon-gamma is species specific, limited by suppressor of cytokine signalling 1 and influences interleukin-17 production. *EMBO Rep.* **2010**, *11*, 640–646. [CrossRef]

36. Eguchi, K.; Manabe, I.; Oishi-Tanaka, Y.; Ohsugi, M.; Kono, N.; Ogata, F.; Yagi, N.; Ohto, U.; Kimoto, M.; Miyake, K.; et al. Saturated fatty acid and TLR signaling link beta cell dysfunction and islet inflammation. *Cell Metab.* **2012**, *15*, 518–533. [CrossRef]

37. Westwell-Roper, C.Y.; Ehses, J.A.; Verchere, C.B. Resident macrophages mediate islet amyloid polypeptide-induced islet IL-1beta production and beta-cell dysfunction. *Diabetes* **2014**, *63*, 1698–1711. [CrossRef]

38. Richardson, S.J.; Willcox, A.; Bone, A.J.; Foulis, A.K.; Morgan, N.G. Islet-associated macrophages in type 2 diabetes. *Diabetologia* **2009**, *52*, 1686–1688. [CrossRef]

39. Kamata, K.; Mizukami, H.; Inaba, W.; Tsuboi, K.; Tateishi, Y.; Yoshida, T.; Yagihashi, S. Islet amyloid with macrophage migration correlates with augmented beta-cell deficits in type 2 diabetic patients. *Amyloid* **2014**, *21*, 191–201. [CrossRef]

40. Collier, J.J.; Sparer, T.E.; Karlstad, M.D.; Burke, S.J. Pancreatic islet inflammation: An emerging role for chemokines. *J. Mol. Endocrinol.* **2017**, *59*, R33–R46. [CrossRef]

41. Talchai, C.; Xuan, S.; Lin, H.V.; Sussel, L.; Accili, D. Pancreatic beta cell dedifferentiation as a mechanism of diabetic beta cell failure. *Cell* **2012**, *150*, 1223–1234. [CrossRef] [PubMed]

42. Chan, J.Y.; Lee, K.; Maxwell, E.L.; Liang, C.; Laybutt, D.R. Macrophage alterations in islets of obese mice linked to beta cell disruption in diabetes. *Diabetologia* **2019**, *62*, 993–999. [CrossRef] [PubMed]

43. He, W.; Yuan, T.; Maedler, K. Macrophage-associated pro-inflammatory state in human islets from obese individuals. *Nutr. Diabetes* **2019**, *9*, 36. [CrossRef] [PubMed]

44. Delmastro, M.M.; Piganelli, J.D. Oxidative stress and redox modulation potential in type 1 diabetes. *Clin. Dev. Immunol.* **2011**, *2011*, 593863. [CrossRef] [PubMed]

45. Burke, S.J.; Batdorf, H.M.; Burk, D.H.; Martin, T.M.; Mendoza, T.; Stadler, K.; Alami, W.; Karlstad, M.D.; Robson, M.J.; Blakely, R.D.; et al. Pancreatic deletion of the interleukin-1 receptor disrupts whole body glucose homeostasis and promotes islet beta-cell de-differentiation. *Mol. Metab.* **2018**, *14*, 95–107. [CrossRef] [PubMed]

46. Burke, S.J.; Stadler, K.; Lu, D.; Gleason, E.; Han, A.; Donohoe, D.R.; Rogers, R.C.; Hermann, G.E.; Karlstad, M.D.; Collier, J.J. IL-1beta reciprocally regulates chemokine and insulin secretion in pancreatic beta-cells via NF-kappaB. *Am. J. Physiol. Endocrinol. Metab.* **2015**, *309*, E715–E726. [CrossRef]

47. Dror, E.; Dalmas, E.; Meier, D.T.; Wueest, S.; Thevenet, J.; Thienel, C.; Timper, K.; Nordmann, T.M.; Traub, S.; Schulze, F.; et al. Postprandial macrophage-derived IL-1beta stimulates insulin, and both synergistically promote glucose disposal and inflammation. *Nat. Immunol.* **2017**, *18*, 283–292. [CrossRef]

48. Welsh, N.; Cnop, M.; Kharroubi, I.; Bugliani, M.; Lupi, R.; Marchetti, P.; Eizirik, D.L. Is there a role for locally produced interleukin-1 in the deleterious effects of high glucose or the type 2 diabetes milieu to human pancreatic islets? *Diabetes* **2005**, *54*, 3238–3244. [CrossRef]

49. Eizirik, D.L.; Mandrup-Poulsen, T. A choice of death—The signal-transduction of immune-mediated beta-cell apoptosis. *Diabetologia* **2001**, *44*, 2115–2133. [CrossRef]

50. Ablamunits, V.; Henegariu, O.; Hansen, J.B.; Opare-Addo, L.; Preston-Hurlburt, P.; Santamaria, P.; Mandrup-Poulsen, T.; Herold, K.C. Synergistic reversal of type 1 diabetes in NOD mice with anti-CD3 and interleukin-1 blockade: Evidence of improved immune regulation. *Diabetes* **2012**, *61*, 145–154. [CrossRef]

51. O'Neill, C.M.; Lu, C.; Corbin, K.L.; Sharma, P.R.; Dula, S.B.; Carter, J.D.; Ramadan, J.W.; Xin, W.; Lee, J.K.; Nunemaker, C.S. Circulating levels of IL-1B+IL-6 cause ER stress and dysfunction in islets from prediabetic male mice. *Endocrinology* **2013**, *154*, 3077–3088. [CrossRef] [PubMed]

52. Burkart, V.; Wang, Z.Q.; Radons, J.; Heller, B.; Herceg, Z.; Stingl, L.; Wagner, E.F.; Kolb, H. Mice lacking the poly(ADP-ribose) polymerase gene are resistant to pancreatic beta-cell destruction and diabetes development induced by streptozocin. *Nat. Med.* **1999**, *5*, 314–319. [CrossRef]

53. Westwell-Roper, C.; Dai, D.L.; Soukhatcheva, G.; Potter, K.J.; van Rooijen, N.; Ehses, J.A.; Verchere, C.B. IL-1 blockade attenuates islet amyloid polypeptide-induced proinflammatory cytokine release and pancreatic islet graft dysfunction. *J. Immunol.* **2011**, *187*, 2755–2765. [CrossRef] [PubMed]

54. Bouzakri, K.; Ribaux, P.; Tomas, A.; Parnaud, G.; Rickenbach, K.; Halban, P.A. Rab GTPase-activating protein AS160 is a major downstream effector of protein kinase B/Akt signaling in pancreatic beta-cells. *Diabetes* **2008**, *57*, 1195–1204. [CrossRef] [PubMed]

55. Zawalich, W.S.; Zawalich, K.C. Interleukin 1 is a potent stimulator of islet insulin secretion and phosphoinositide hydrolysis. *Am. J. Physiol.* **1989**, *256*, E19–E24. [CrossRef] [PubMed]

56. Cucak, H.; Mayer, C.; Tonnesen, M.; Thomsen, L.H.; Grunnet, L.G.; Rosendahl, A. Macrophage Contact Dependent and Independent TLR4 Mechanisms Induce beta-Cell Dysfunction and Apoptosis in a Mouse Model of Type 2 Diabetes. *PLoS ONE* **2014**, *9*. [CrossRef]

57. Zhou, R.; Tardivel, A.; Thorens, B.; Choi, I.; Tschopp, J. Thioredoxin-interacting protein links oxidative stress to inflammasome activation. *Nat. Immunol.* **2010**, *11*, 136–140. [CrossRef]

58. Ehses, J.A.; Lacraz, G.; Giroix, M.H.; Schmidlin, F.; Coulaud, J.; Kassis, N.; Irminger, J.C.; Kergoat, M.; Portha, B.; Homo-Delarche, F.; et al. IL-1 antagonism reduces hyperglycemia and tissue inflammation in the type 2 diabetic GK rat. *Proc. Natl. Acad. Sci. USA* **2009**, *106*, 13998–14003. [CrossRef]

59. Sauter, N.S.; Thienel, C.; Plutino, Y.; Kampe, K.; Dror, E.; Traub, S.; Timper, K.; Bedat, B.; Pattou, F.; Kerr-Conte, J.; et al. Angiotensin II induces interleukin-1beta-mediated islet inflammation and beta-cell dysfunction independently of vasoconstrictive effects. *Diabetes* **2015**, *64*, 1273–1283. [CrossRef]

60. Novak, I.; Solini, A. P2X receptor-ion channels in the inflammatory response in adipose tissue and pancreas-potential triggers in onset of type 2 diabetes? *Curr. Opin. Immunol.* **2018**, *52*, 1–7. [CrossRef]

61. Cnop, M.; Abdulkarim, B.; Bottu, G.; Cunha, D.A.; Igoillo-Esteve, M.; Masini, M.; Turatsinze, J.V.; Griebel, T.; Villate, O.; Santin, I.; et al. RNA sequencing identifies dysregulation of the human pancreatic islet transcriptome by the saturated fatty acid palmitate. *Diabetes* **2014**, *63*, 1978–1993. [CrossRef] [PubMed]

62. Diana, J.; Lehuen, A. Macrophages and beta-cells are responsible for CXCR2-mediated neutrophil infiltration of the pancreas during autoimmune diabetes. *EMBO Mol. Med.* **2014**, *6*, 1090–1104. [CrossRef]

63. Eizirik, D.L.; Sammeth, M.; Bouckenooghe, T.; Bottu, G.; Sisino, G.; Igoillo-Esteve, M.; Ortis, F.; Santin, I.; Colli, M.L.; Barthson, J.; et al. The human pancreatic islet transcriptome: Expression of candidate genes for type 1 diabetes and the impact of pro-inflammatory cytokines. *PLoS Genet.* **2012**, *8*, e1002552. [CrossRef]

64. Ward, M.G.; Li, G.; Hao, M. Apoptotic beta-cells induce macrophage reprogramming under diabetic conditions. *J. Biol. Chem.* **2018**, *293*, 16160–16173. [CrossRef]

65. Riera-Borrull, M.; Cuevas, V.D.; Alonso, B.; Vega, M.A.; Joven, J.; Izquierdo, E.; Corbi, A.L. Palmitate Conditions Macrophages for Enhanced Responses toward Inflammatory Stimuli via JNK Activation. *J. Immunol.* **2017**, *199*, 3858–3869. [CrossRef]

66. Hajmrle, C.; Smith, N.; Spigelman, A.F.; Dai, X.; Senior, L.; Bautista, A.; Ferdaoussi, M.; MacDonald, P.E. Interleukin-1 signaling contributes to acute islet compensation. *JCI Insight* **2016**, *1*, e86055. [CrossRef]

67. Zawalich, W.S.; Dierolf, B.; Zawalich, K.C. Interleukin-1 induces time-dependent potentiation in isolated rat islets: Possible involvement of phosphoinositide hydrolysis. *Endocrinology* **1989**, *124*, 720–726. [CrossRef]

68. Dasu, M.R.; Devaraj, S.; Jialal, I. High glucose induces IL-1beta expression in human monocytes: Mechanistic insights. *Am. J. Physiol. Endocrinol. Metab.* **2007**, *293*, E337–E346. [CrossRef]

69. Nordmann, T.M.; Dror, E.; Schulze, F.; Traub, S.; Berishvili, E.; Barbieux, C.; Boni-Schnetzler, M.; Donath, M.Y. The Role of Inflammation in beta-cell Dedifferentiation. *Sci. Rep.* **2017**, *7*, 6285. [CrossRef]

70. Oshima, M.; Knoch, K.P.; Diedisheim, M.; Petzold, A.; Cattan, P.; Bugliani, M.; Marchetti, P.; Choudhary, P.; Huang, G.C.; Bornstein, S.R.; et al. Virus-like infection induces human beta cell dedifferentiation. *JCI Insight* **2018**, *3*. [CrossRef]

71. Ibarra Urizar, A.; Prause, M.; Wortham, M.; Sui, Y.; Thams, P.; Sander, M.; Christensen, G.L.; Billestrup, N. Beta-cell dysfunction induced by non-cytotoxic concentrations of Interleukin-1beta is associated with changes in expression of beta-cell maturity genes and associated histone modifications. *Mol. Cell. Endocrinol.* **2019**, *496*, 110524. [CrossRef] [PubMed]

72. Kaneto, H.; Xu, G.; Fujii, N.; Kim, S.; Bonner-Weir, S.; Weir, G.C. Involvement of c-Jun N-terminal kinase in oxidative stress-mediated suppression of insulin gene expression. *J. Biol. Chem.* **2002**, *277*, 30010–30018. [CrossRef] [PubMed]

73. Spinas, G.A.; Hansen, B.S.; Linde, S.; Kastern, W.; Molvig, J.; Mandruppoulsen, T.; Dinarello, C.A.; Nielsen, J.H.; Nerup, J. Interleukin-1 Dose-Dependently Affects the Biosynthesis of (Pro)Insulin in Isolated Rat Islets of Langerhans. *Diabetologia* **1987**, *30*, 474–480. [CrossRef] [PubMed]

74. Tanaka, T.; Narazaki, M.; Kishimoto, T. IL-6 in inflammation, immunity, and disease. *Cold Spring Harb. Perspect. Biol.* **2014**, *6*, a016295. [CrossRef]

75. Kristiansen, O.P.; Mandrup-Poulsen, T. Interleukin-6 and diabetes: The good, the bad, or the indifferent? *Diabetes* **2005**, *54*, S114–S124. [CrossRef]

76. Brozzi, F.; Nardelli, T.R.; Lopes, M.; Millard, I.; Barthson, J.; Igoillo-Esteve, M.; Grieco, F.A.; Villate, O.; Oliveira, J.M.; Casimir, M.; et al. Cytokines induce endoplasmic reticulum stress in human, rat and mouse beta cells via different mechanisms. *Diabetologia* **2015**, *58*, 2307–2316. [CrossRef]

77. Maedler, K.; Spinas, G.A.; Lehmann, R.; Sergeev, P.; Weber, M.; Fontana, A.; Kaiser, N.; Donath, M.Y. Glucose induces beta-cell apoptosis via upregulation of the Fas receptor in human islets. *Diabetes* **2001**, *50*, 1683–1690. [CrossRef]

78. He, W.; Rebello, O.; Savino, R.; Terracciano, R.; Schuster-Klein, C.; Guardiola, B.; Maedler, K. TLR4 triggered complex inflammation in human pancreatic islets. *Biochim. Biophys. Acta Mol. Basis. Dis.* **2019**, *1865*, 86–97. [CrossRef]

79. Nackiewicz, D.; Dan, M.; He, W.; Kim, R.; Salmi, A.; Rutti, S.; Westwell-Roper, C.; Cunningham, A.; Speck, M.; Schuster-Klein, C.; et al. TLR2/6 and TLR4-activated macrophages contribute to islet inflammation and impair beta cell insulin gene expression via IL-1 and IL-6. *Diabetologia* **2014**, *57*, 1645–1654. [CrossRef]

80. Baltrusch, S. Mitochondrial network regulation and its potential interference with inflammatory signals in pancreatic beta cells. *Diabetologia* **2016**, *59*, 683–687. [CrossRef]

81. Zheng, Q.Y.; Cao, Z.H.; Hu, X.B.; Li, G.Q.; Dong, S.F.; Xu, G.L.; Zhang, K.Q. LIGHT/IFN-gamma triggers beta cells apoptosis via NF-kappaB/Bcl2-dependent mitochondrial pathway. *J. Cell Mol. Med.* **2016**, *20*, 1861–1871. [CrossRef]

82. Cao, Z.H.; -Zheng, Q.Y.; Li, G.Q.; Hu, X.B.; Feng, S.L.; Xu, G.L.; Zhang, K.Q. STAT1-mediated down-regulation of Bcl-2 expression is involved in IFN-gamma/TNF-alpha-induced apoptosis in NIT-1 cells. *PLoS ONE* **2015**, *10*, e0120921. [CrossRef]

83. Gysemans, C.A.; Ladriere, L.; Callewaert, H.; Rasschaert, J.; Flamez, D.; Levy, D.E.; Matthys, P.; Eizirik, D.L.; Mathieu, C. Disruption of the gamma-interferon signaling pathway at the level of signal transducer and activator of transcription-1 prevents immune destruction of beta-cells. *Diabetes* **2005**, *54*, 2396–2403. [CrossRef]

84. Baldeon, M.E.; Chun, T.; Gaskins, H.R. Interferon-alpha and interferon-gamma differentially affect pancreatic beta-cell phenotype and function. *Am. J. Physiol.* **1998**, *275*, C25–C32. [CrossRef]

85. Freudenburg, W.; Gautam, M.; Chakraborty, P.; James, J.; Richards, J.; Salvatori, A.S.; Baldwin, A.; Schriewer, J.; Buller, R.M.L.; Corbett, J.A.; et al. Reduction in ATP Levels Triggers Immunoproteasome Activation by the 11S (PA28) Regulator during Early Antiviral Response Mediated by IFN beta in Mouse Pancreatic beta-Cells. *PLoS ONE* **2013**, *8*, e52408. [CrossRef]

86. Heitmeier, M.R.; Scarim, A.L.; Corbett, J.A. Interferon-gamma increases the sensitivity of islets of Langerhans for inducible nitric-oxide synthase expression induced by interleukin 1. *J. Biol. Chem.* **1997**, *272*, 13697–13704. [CrossRef]

87. Oleson, B.J.; Corbett, J.A. Dual Role of Nitric Oxide in Regulating the Response of beta Cells to DNA Damage. *Antioxid. Redox Sign.* **2018**, *29*, 1432–1445. [CrossRef]

88. Hughes, K.J.; Meares, G.P.; Hansen, P.A.; Corbett, J.A. FoxO1 and SIRT1 Regulate beta-Cell Responses to Nitric Oxide. *J. Biol. Chem.* **2011**, *286*, 8338–8348. [CrossRef]

89. Dahlen, E.; Dawe, K.; Ohlsson, L.; Hedlund, G. Dendritic cells and macrophages are the first and major producers of TNF-alpha in pancreatic islets in the nonobese diabetic mouse. *J. Immunol.* **1998**, *160*, 3585–3593.

90. Odegaard, J.I.; Chawla, A. Connecting Type 1 and Type 2 Diabetes through Innate Immunity. *Cold Spring Harb. Perspect. Med.* **2012**, *2*. [CrossRef]

91. Janjuha, S.; Singh, S.P.; Tsakmaki, A.; Mousavy Gharavy, S.N.; Murawala, P.; Konantz, J.; Birke, S.; Hodson, D.J.; Rutter, G.A.; Bewick, G.A.; et al. Age-related islet inflammation marks the proliferative decline of pancreatic beta-cells in zebrafish. *eLife* **2018**, *7*. [CrossRef] [PubMed]

92. Turner, D.A.; Paszek, P.; Woodcock, D.J.; Nelson, D.E.; Horton, C.A.; Wang, Y.; Spiller, D.G.; Rand, D.A.; White, M.R.; Harper, C.V. Physiological levels of TNFalpha stimulation induce stochastic dynamics of NF-kappaB responses in single living cells. *J. Cell Sci.* **2010**, *123*, 2834–2843. [CrossRef]

93. Eguchi, K.; Manabe, I. Macrophages and islet inflammation in type 2 diabetes. *Diabetes Obes. Metab.* **2013**, *15* (Suppl. 3), 152–158. [CrossRef]

94. Morgan, D.; Oliveira-Emilio, H.R.; Keane, D.; Hirata, A.E.; Santos da Rocha, M.; Bordin, S.; Curi, R.; Newsholme, P.; Carpinelli, A.R. Glucose, palmitate and pro-inflammatory cytokines modulate production and activity of a phagocyte-like NADPH oxidase in rat pancreatic islets and a clonal beta cell line. *Diabetologia* **2007**, *50*, 359–369. [CrossRef] [PubMed]

95. Eizirik, D.L.; Flodstrom, M.; Karlsen, A.E.; Welsh, N. The harmony of the spheres: Inducible nitric oxide synthase and related genes in pancreatic beta cells. *Diabetologia* **1996**, *39*, 875–890. [CrossRef] [PubMed]

96. Quintana-Lopez, L.; Blandino-Rosano, M.; Perez-Arana, G.; Cebada-Aleu, A.; Lechuga-Sancho, A.; Aguilar-Diosdado, M.; Segundo, C. Nitric Oxide Is a Mediator of Antiproliferative Effects Induced by Proinflammatory Cytokines on Pancreatic Beta Cells. *Mediat. Inflamm.* **2013**, *2013*. [CrossRef] [PubMed]

97. Grunnet, L.G.; Aikin, R.; Tonnesen, M.F.; Paraskevas, S.; Blaabjerg, L.; Storling, J.; Rosenberg, L.; Billestrup, N.; Maysinger, D.; Mandrup-Poulsen, T. Proinflammatory cytokines activate the intrinsic apoptotic pathway in beta-cells. *Diabetes* **2009**, *58*, 1807–1815. [CrossRef]

98. Pavlovic, D.; Chen, M.C.; Gysemans, C.A.; Mathieu, C.; Eizirik, D.L. The role of interferon regulatory factor-1 in cytokine-induced mRNA expression and cell death in murine pancreatic beta-cells. *Eur. Cytokine Netw.* **1999**, *10*, 403–412.

99. Gysemans, C.; Callewaert, H.; Moore, F.; Nelson-Holte, M.; Overbergh, L.; Eizirik, D.L.; Mathieu, C. Interferon regulatory factor-1 is a key transcription factor in murine beta cells under immune attack. *Diabetologia* **2009**, *52*, 2374–2384. [CrossRef]

100. Grieco, F.A.; Moore, F.; Vigneron, F.; Santin, I.; Villate, O.; Marselli, L.; Rondas, D.; Korf, H.; Overbergh, L.; Dotta, F.; et al. IL-17A increases the expression of proinflammatory chemokines in human pancreatic islets. *Diabetologia* **2014**, *57*, 502–511. [CrossRef]

101. Parkash, J.; Chaudhry, M.A.; Rhoten, W.B. Tumor necrosis factor-alpha-induced changes in insulin-producing beta-cells. *Anat. Rec. A Discov. Mol. Cell Evol. Biol.* **2005**, *286*, 982–993. [CrossRef] [PubMed]

102. Geutskens, S.B.; Otonkoski, T.; Pulkkinen, M.A.; Drexhage, H.A.; Leenen, P.J. Macrophages in the murine pancreas and their involvement in fetal endocrine development in vitro. *J. Leukoc. Biol.* **2005**, *78*, 845–852. [CrossRef] [PubMed]

103. Banaei-Bouchareb, L.; Gouon-Evans, V.; Samara-Boustani, D.; Castellotti, M.C.; Czernichow, P.; Pollard, J.W.; Polak, M. Insulin cell mass is altered in Csf1op/Csf1op macrophage-deficient mice. *J. Leukoc. Biol.* **2004**, *76*, 359–367. [CrossRef] [PubMed]

104. Banaei-Bouchareb, L.; Peuchmaur, M.; Czernichow, P.; Polak, M. A transient microenvironment loaded mainly with macrophages in the early developing human pancreas. *J. Endocrinol.* **2006**, *188*, 467–480. [CrossRef] [PubMed]

105. Mosser, R.E.; Maulis, M.F.; Moulle, V.S.; Dunn, J.C.; Carboneau, B.A.; Arasi, K.; Pappan, K.; Poitout, V.; Gannon, M. High-fat diet-induced beta-cell proliferation occurs prior to insulin resistance in C57Bl/6J male mice. *Am. J. Physiol. Endocrinol. Metab.* **2015**, *308*, E573–E582. [CrossRef]

106. Woodland, D.C.; Liu, W.; Leong, J.; Sears, M.L.; Luo, P.; Chen, X. Short-term high-fat feeding induces islet macrophage infiltration and beta-cell replication independently of insulin resistance in mice. *Am. J. Physiol. Endocrinol. Metab.* **2016**, *311*, E763–E771. [CrossRef]

107. Criscimanna, A.; Coudriet, G.M.; Gittes, G.K.; Piganelli, J.D.; Esni, F. Activated macrophages create lineage-specific microenvironments for pancreatic acinar- and beta-cell regeneration in mice. *Gastroenterology* **2014**, *147*, 1106–1118.e1111. [CrossRef]

108. Riley, K.G.; Pasek, R.C.; Maulis, M.F.; Dunn, J.C.; Bolus, W.R.; Kendall, P.L.; Hasty, A.H.; Gannon, M. Macrophages are essential for CTGF-mediated adult beta-cell proliferation after injury. *Mol. Metab.* **2015**, *4*, 584–591. [CrossRef]

109. Xiao, X.; Guo, P.; Prasadan, K.; Shiota, C.; Peirish, L.; Fischbach, S.; Song, Z.; Gaffar, I.; Wiersch, J.; El-Gohary, Y.; et al. Pancreatic cell tracing, lineage tagging and targeted genetic manipulations in multiple cell types using pancreatic ductal infusion of adeno-associated viral vectors and/or cell-tagging dyes. *Nat. Protoc.* **2014**, *9*, 2719–2724. [CrossRef]

110. Guo, P.; Preuett, B.; Krishna, P.; Xiao, X.; Shiota, C.; Wiersch, J.; Gaffar, I.; Tulachan, S.; El-Gohary, Y.; Song, Z.; et al. Barrier function of the coelomic epithelium in the developing pancreas. *Mech. Dev.* **2014**, *134*, 67–79. [CrossRef]

111. Xiao, X.; Gaffar, I.; Guo, P.; Wiersch, J.; Fischbach, S.; Peirish, L.; Song, Z.; El-Gohary, Y.; Prasadan, K.; Shiota, C.; et al. M2 macrophages promote beta-cell proliferation by up-regulation of SMAD7. *Proc. Natl. Acad. Sci. USA* **2014**, *111*, E1211–E1220. [CrossRef] [PubMed]

112. Xiao, X.; Prasadan, K.; Guo, P.; El-Gohary, Y.; Fischbach, S.; Wiersch, J.; Gaffar, I.; Shiota, C.; Gittes, G.K. Pancreatic duct cells as a source of VEGF in mice. *Diabetologia* **2014**, *57*, 991–1000. [CrossRef] [PubMed]

113. Tessem, J.S.; Jensen, J.N.; Pelli, H.; Dai, X.M.; Zong, X.H.; Stanley, E.R.; Jensen, J.; DeGregori, J. Critical roles for macrophages in islet angiogenesis and maintenance during pancreatic degeneration. *Diabetes* **2008**, *57*, 1605–1617. [CrossRef] [PubMed]

114. Reinert, R.B.; Cai, Q.; Hong, J.Y.; Plank, J.L.; Aamodt, K.; Prasad, N.; Aramandla, R.; Dai, C.; Levy, S.E.; Pozzi, A.; et al. Vascular endothelial growth factor coordinates islet innervation via vascular scaffolding. *Development* **2014**, *141*, 1480–1491. [CrossRef] [PubMed]

115. Brissova, M.; Aamodt, K.; Brahmachary, P.; Prasad, N.; Hong, J.Y.; Dai, C.; Mellati, M.; Shostak, A.; Poffenberger, G.; Aramandla, R.; et al. Islet microenvironment, modulated by vascular endothelial growth factor-A signaling, promotes beta cell regeneration. *Cell Metab.* **2014**, *19*, 498–511. [CrossRef]

116. Smeets, S.; Stange, G.; Leuckx, G.; Roelants, L.; Cools, W.; De Paep, D.L.; Ling, Z.; De Leu, N.; In't Veld, P. Evidence of Tissue Repair in Human Donor Pancreas After Prolonged Duration of Stay in Intensive Care. *Diabetes* **2020**, *69*, 401–412. [CrossRef]

117. Cao, X.; Han, Z.B.; Zhao, H.; Liu, Q. Transplantation of mesenchymal stem cells recruits trophic macrophages to induce pancreatic beta cell regeneration in diabetic mice. *Int. J. Biochem. Cell Biol.* **2014**, *53*, 372–379. [CrossRef]

118. Criscimanna, A.; Speicher, J.A.; Houshmand, G.; Shiota, C.; Prasadan, K.; Ji, B.; Logsdon, C.D.; Gittes, G.K.; Esni, F. Duct cells contribute to regeneration of endocrine and acinar cells following pancreatic damage in adult mice. *Gastroenterology* **2011**, *141*, 1451–1462.e6. [CrossRef]

119. Dalmas, E.; Lehmann, F.M.; Dror, E.; Wueest, S.; Thienel, C.; Borsigova, M.; Stawiski, M.; Traunecker, E.; Lucchini, F.C.; Dapito, D.H.; et al. Interleukin-33-Activated Islet-Resident Innate Lymphoid Cells Promote Insulin Secretion through Myeloid Cell Retinoic Acid Production. *Immunity* **2017**, *47*, 928–942.e927. [CrossRef]

120. El-Gohary, Y.; Tulachan, S.; Wiersch, J.; Guo, P.; Welsh, C.; Prasadan, K.; Paredes, J.; Shiota, C.; Xiao, X.; Wada, Y.; et al. A smad signaling network regulates islet cell proliferation. *Diabetes* **2014**, *63*, 224–236. [CrossRef]

121. El-Gohary, Y.; Tulachan, S.; Guo, P.; Welsh, C.; Wiersch, J.; Prasadan, K.; Paredes, J.; Shiota, C.; Xiao, X.; Wada, Y.; et al. Smad signaling pathways regulate pancreatic endocrine development. *Dev. Biol.* **2013**, *378*, 83–93. [CrossRef] [PubMed]

122. Ying, W.; Lee, Y.S.; Dong, Y.; Seidman, J.S.; Yang, M.; Isaac, R.; Seo, J.B.; Yang, B.H.; Wollam, J.; Riopel, M.; et al. Expansion of Islet-Resident Macrophages Leads to Inflammation Affecting beta Cell Proliferation and Function in Obesity. *Cell Metab.* **2019**, *29*, 457–474.e5. [CrossRef] [PubMed]

123. Nishikoba, N.; Kumagai, K.; Kanmura, S.; Nakamura, Y.; Ono, M.; Eguchi, H.; Kamibayashiyama, T.; Oda, K.; Mawatari, S.; Tanoue, S.; et al. HGF-MET Signaling Shifts M1 Macrophages Toward an M2-Like Phenotype Through PI3K-Mediated Induction of Arginase-1 Expression. *Fron. Immunol.* **2020**, *11*. [CrossRef] [PubMed]

124. Garcia-Ocana, A.; Takane, K.K.; Syed, M.A.; Philbrick, W.M.; Vasavada, R.C.; Stewart, A.F. Hepatocyte growth factor overexpression in the islet of transgenic mice increases beta cell proliferation, enhances islet mass, and induces mild hypoglycemia. *J. Biol. Chem.* **2000**, *275*, 1226–1232. [CrossRef] [PubMed]

125. Alvarez-Perez, J.C.; Ernst, S.; Demirci, C.; Casinelli, G.P.; Mellado-Gil, J.M.D.; Rausell-Palamos, F.; Vasavada, R.C.; Garcia-Ocana, A. Hepatocyte Growth Factor/c-Met Signaling Is Required for beta-Cell Regeneration. *Diabetes* **2014**, *63*, 216–223. [CrossRef] [PubMed]

126. Nackiewicz, D.; Dan, M.X.; Speck, M.; Chow, S.Z.; Chen, Y.C.; Pospisilik, J.A.; Verchere, C.B.; Ehses, J.A. Islet Macrophages Shift to a Reparative State following Pancreatic Beta-Cell Death and Area Major Source of Islet Insulin-like Growth Factor-1. *Iscience* **2020**, *23*. [CrossRef]

127. Kaminski, A.; Kaminski, E.R.; Morgan, N.G. Pre-incubation with interleukin-4 mediates a direct protective effect against the loss of pancreatic beta-cell viability induced by proinflammatory cytokines. *Clin. Exp. Immunol.* **2007**, *148*, 583–588. [CrossRef]

128. Souza, K.L.A.; Gurgul-Convey, E.; Elsner, M.; Lenzen, S. Interaction between pro-inflammatory and anti-inflammatory cytokines in insulin-producing cells. *J. Endocrinol.* **2008**, *197*, 139–150. [CrossRef]

129. Pfleger, C.; Meierhoff, G.; Kolb, H.; Schloot, N.C.; Grp, P.S. Association of T-cell reactivity with beta-cell function in recent onset type 1 diabetes patients. *J. Autoimmun.* **2010**, *34*, 127–135. [CrossRef]

130. Weitz, J.R.; Jacques-Silva, C.; Qadir, M.M.F.; Umland, O.; Pereira, E.; Qureshi, F.; Tamayo, A.; Dominguez-Bendala, J.; Rodriguez-Diaz, R.; Almaca, J.; et al. Secretory Functions of Macrophages in the Human Pancreatic Islet Are Regulated by Endogenous Purinergic Signaling. *Diabetes* **2020**, *69*, 1206–1218. [CrossRef]

131. Fu, W.; Farache, J.; Clardy, S.M.; Hattori, K.; Mander, P.; Lee, K.; Rioja, I.; Weissleder, R.; Prinjha, R.K.; Benoist, C.; et al. Epigenetic modulation of type-1 diabetes via a dual effect on pancreatic macrophages and beta cells. *eLife* **2014**, *3*, e04631. [CrossRef] [PubMed]

132. Parsa, R.; Andresen, P.; Gillett, A.; Mia, S.; Zhang, X.M.; Mayans, S.; Holmberg, D.; Harris, R.A. Adoptive transfer of immunomodulatory M2 macrophages prevents type 1 diabetes in NOD mice. *Diabetes* **2012**, *61*, 2881–2892. [CrossRef] [PubMed]

133. Zheng, D.; Wang, Y.; Cao, Q.; Lee, V.W.; Zheng, G.; Sun, Y.; Tan, T.K.; Wang, Y.; Alexander, S.I.; Harris, D.C. Transfused macrophages ameliorate pancreatic and renal injury in murine diabetes mellitus. *Nephron. Exp. Nephrol.* **2011**, *118*, e87–e99. [CrossRef] [PubMed]

Publisher's Note: MDPI stays neutral with regard to jurisdictional claims in published maps and institutional affiliations.

Review

Oxidative Stress in Cytokine-Induced Dysfunction of the Pancreatic Beta Cell: Known Knowns and Known Unknowns

Anjaneyulu Kowluru

Biomedical Research Service, John D. Dingell VA Medical Center, Department of Pharmaceutical Sciences, Eugene Applebaum College of Pharmacy and Health Sciences, Wayne State University, Detroit, MI 48201, USA; akowluru@med.wayne.edu

Received: 20 October 2020; Accepted: 21 November 2020; Published: 24 November 2020

Abstract: Compelling evidence from earlier studies suggests that the pancreatic beta cell is inherently weak in its antioxidant defense mechanisms to face the burden of protecting itself against the increased intracellular oxidative stress following exposure to proinflammatory cytokines. Recent evidence implicates novel roles for nicotinamide adenine dinucleotide phosphate (NADPH) oxidases (Noxs) as contributors to the excessive intracellular oxidative stress and damage under metabolic stress conditions. This review highlights the existing evidence on the regulatory roles of at least three forms of Noxs, namely Nox1, Nox2, and Nox4, in the cascade of events leading to islet beta cell dysfunction, specifically under the duress of chronic exposure to cytokines. Potential crosstalk between key signaling pathways (e.g., inducible nitric oxide synthase [iNOS] and Noxs) in the generation and propagation of reactive molecules and metabolites leading to mitochondrial damage and cell apoptosis is discussed. Available data accrued in investigations involving small-molecule inhibitors and antioxidant protein expression methods as tools toward the prevention of cytokine-induced oxidative damage are reviewed. Lastly, current knowledge gaps in this field, and possible avenues for future research are highlighted.

Keywords: proinflammatory cytokines; oxidative stress; NADPH oxidases; Rac1; pancreatic beta cell; diabetes

---

1. Introduction

Type-1 diabetes (T1DM) is characterized by an absolute insulin deficiency arising from autoimmune destruction of the pancreatic islet β-cell. It is generally accepted that, during the course of the progression of T1DM, proinflammatory cytokines (e.g., IL-1β, TNF-α, and IFN-γ) are released by infiltrating activated immune cells. However, the exact molecular and cellular mechanisms by which cytokines induce β-cell dysregulation and demise remain only partially understood. In this context, while a host of competing signaling pathways have been proposed to contribute to the beta cell dysfunction under the duress of cytokines, apoptosis is considered as the primary mode of beta cell death in human and mouse models of T1DM [1–3]. Interestingly, published evidence also implicates mitochondrial dysfunction as the hallmark of beta cell demise under exposure to proinflammatory cytokines [4–9]. The mitochondrial damage via loss of membrane permeability pore transition (MMPT) leads to the release of the mitochondrial cytochrome C into the cytosolic compartment to promote the activation of caspases, culminating in the degradation and functional inactivation of key intracellular proteins that may be necessary for cell proliferation and survival, including G protein prenylating enzymes and nuclear lamins [10,11]. From a mechanistic standpoint, extant studies have suggested potential roles of increased oxidative stress, inflammation, and endoplasmic reticulum (ER) stress as contributors to the islet beta cell dysfunction under the duress of the above-mentioned pathological stimuli. The reader is referred to select reviews highlighting the contributory roles of increased intracellular metabolic stress in the pathology of islet beta-cell destruction under diabetogenic conditions [2,5,12–15].

At the outset, it is important to note that the pancreatic beta cell is relatively more vulnerable to oxidative damage due to the inherent deficiency of a strong antioxidant capacity to counteract the excessive generation of reactive oxygen species (ROS) under conditions of exposure to various pathological insults, including exposure to proinflammatory cytokines. In this context, original contributions from the laboratory of Lenzen and coworkers provided compelling evidence demonstrating poor antioxidant enzymatic machinery in the islet beta cell [16]. Using Northern blot hybridization methodology, these investigators quantified the gene expression of various antioxidant enzymes in mouse tissues. Their data revealed significantly low levels of these genes in pancreatic islets compared to the other tissues studied. For example, Cu-Zn superoxide dismutase and Mn-superoxide dismutase (Mn-SOD) activities in the islet were only 38% and 30%, respectively, of the levels of these enzymes in the liver. Likewise, the expression of the glutathione peroxidase (GPx) gene in the islet was only 15% of the level seen in the liver. Lastly, catalase gene expression was undetectable in the islet. Based on the above findings, these investigators proposed that a relatively low abundance of the antioxidant enzymes in the islet may contribute to its susceptibility to oxidative stress in human and animal diabetes [16]. Several follow-up studies assessed the antioxidant capacity of human islet beta cells. Gurgul-Convey and coworkers demonstrated that the antioxidant enzyme profiles in clonal EndoC-βH1 human beta cells are comparable to those in human and rodent islets. Specifically, these studies have shown relatively high levels of SODs and low levels of H_2O_2-inactivating GPx and catalase in these cells [17]. Along these lines, more recent studies by Miki and coworkers [18] demonstrated a significantly high degree of susceptibility of beta cells to oxidative stress compared to alpha cells in the human islet; this is further supported by remarkably low expression levels of GPx and catalase in human beta cells compared to the alpha cells. Together, data accrued in these investigations indicate a clear imbalance between the enzymatic machinery involved in the generation and removal of H_2O_2 in pancreatic beta cells in critical intracellular compartments, including mitochondria, leading to cellular dysregulation and demise under the duress of proinflammatory cytokines [17,19,20]. It is well established that IL-1β mediates its cytotoxic effects on rat beta cells via accelerating the NFκB-mediated induction of the inducible nitric oxide synthase (iNOS) and the release of nitric oxide (NO) and downstream signaling events, culminating in cell dysfunction. Interestingly, however, studies in clonal EndoC-βH1 human beta cells, human islets, and mouse islets (not rat islets) implicate NO-independent effects of cytokines in the induction of metabolic stress and cellular dysregulation [17,21–23]. These observations clearly provide additional support and clarification for the differences reported in earlier studies with regard to differential responses and effects of proinflammatory cytokines on rat, mouse, and human islets and a wide variety of clonal beta cell lines [21]. Based on the above discussion, a clear picture is emerging to suggest critical roles for an inefficient handling of high levels of intracellularly generated H_2O_2 in the beta cell under the duress of cytokines, leading to the pathology of islet beta cell dysfunction and demise. Altogether, a host of intracellularly generated reactive oxygen species (ROS, e.g., superoxide radicals, hydroxyl radicals, and H_2O_2) and reactive nitrogen species (RNS, e.g., NO and peroxynitrite) could play significant roles in the induction of proinflammatory cytokine-induced damage to the islet beta cell [20]. Lastly, superoxide radicals, which are generated by NADPH oxidases (Noxs), undergo dismutation by SODs to promote generation of H_2O_2 in relevant subcellular compartments (e.g., cytosol) in the cytokine-challenged beta cell, leading to damage and loss of functional beta cell mass (see below). The reader is referred to previously published articles for additional details highlighting the deficiency of a robust antioxidant status and the activation of NADPH oxidases as the potential contributing factors for islet beta cell failure and demise under the duress of diabetogenic conditions [20,24–26].

2. NADPH Oxidases Play Key Roles in Islet Function in Health and under Stress Induced by Cytokines

In the back drop of the above discussion about the unique situation the islet beta cell faces, namely poor antioxidant defense, and a high degree of intracellular oxidative stress, which is created under diabetogenic conditions, I have overviewed our current understanding of the roles of NADPH oxidases (Noxs) in the onset of metabolic dysfunction of the beta cell under the duress of exposure to pro-inflammatory cytokines. The Nox superfamily represents a class of flavocytochromes that promote transport electrons through biological membranes and catalyze the cytosolic NADPH-dependent reduction of molecular oxygen to superoxide radicals [25–35]. Interestingly, however, despite a considerable degree of similarity in their ability to generate high levels of superoxide radicals under metabolic stress conditions, they significantly differ in structural composition, subcellular distribution, and response to specific external stimuli (Figure 1). Briefly, the Nox superfamily consists of seven members, namely Nox1-5, dual oxidases 1 (Duox1), and 2 (Duox2). Nox1-3 are membrane bound and require other cytosolic core proteins for holoenzyme assembly and activation. For example, the regulatory components for Nox1 include p22^{phox}, NOX organizers NOXO1 and NOXA2, and the small G protein Rac1. The Nox2 holoenzyme is comprised of membranous cytochrome b558, a heterodimer consisting of p22^{phox}, gp91^{phox}, and the cytosolic core of proteins, including p40^{phox}, p47^{phox}, p67^{phox}, and small G protein Rac1. It has also been proposed that Rap1, a membrane-associated small G protein, contributes to functional regulation of Nox2 holoenzyme [36]. However, potential regulatory roles of Rap1 in functional activation of Nox2 or other forms of Nox in the pancreatic beta cell remain to be elucidated further.

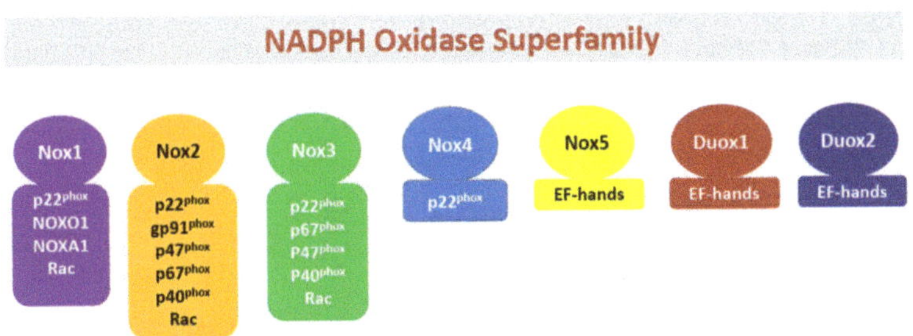

Figure 1. A schematic representation of various members and their subunit composition of the Nox superfamily.

The Nox3 is comprised of p22^{phox}, NOXO1, NOXA1, and Rac1. Interestingly, Nox4, which is localized intracellularly, requires only p22^{phox} but no cytosolic core of proteins. From a mechanistic standpoint, its activity is regulated by its expression, and hence, it is considered constitutively active [29]. It is important to note that Nox 1-4 have a critical requirement for p22^{phox} and Nox1-3 require Rac1 for optimal catalytic function [33]. Lastly, Nox5, Duox1, and Duox2 are associated with the plasma membrane and do not require the intermediacy of cytosolic core of proteins [29,33]. However, these Noxs have calcium-binding motifs (EF hands) for optimal activation (Figure 1). Lastly, it is noteworthy that the Nox5 gene is not expressed in mice and rats [32]. The reader is referred to recent reviews that highlight subunit composition, regulation of individual subunit function via post-translational modifications, translocation to the membrane, holoenzyme assembly, and functional activation of individual members of the Nox superfamily [25–34]. It may be germane to point out that investigations from multiple laboratories have reported expression of Nox1 [37–39], Nox2 [26,40–43], Nox4 [41,44], and Nox5 [45] in human islets, thus making these oxidases potential regulators of islet function and dysfunction in health and disease. Lastly, it should be noted that, of the seven members that belong to the Nox superfamily, only Nox1 and Nox2 are studied extensively in the context of their roles in cytokine-mediated dysregulation of the islet beta cell (see below). Therefore, this review will focus on potential roles of Nox1 and Nox2 in the onset of metabolic defects in the pancreatic islet beta cell under the duress of proinflammatory cytokines.

3. Regulatory Roles of Noxs in Physiological Insulin Secretion

Published evidence from several laboratories has provided convincing evidence on the expression of various members of the Nox superfamily, specifically Nox1, Nox2, and Nox4, in pancreatic beta cells. Examples of relevant studies are highlighted here. Using a variety of complementary experimental approaches, Rebeleto and coworkers [40] reported expression of various subunits of Nox2 in clonal pancreatic beta cells, and rodent and human islets. In addition, they demonstrated inhibition of glucose metabolism and GSIS in these cells following inhibition of Nox2, thus suggesting regulatory roles of Nox2 in physiological insulin secretion. Along these lines, studies by Uchizono and Sumimoto [46] showed an association of Nox1, Nox2, Nox4, and p22^{phox} with the membrane in rodent pancreatic islets. Furthermore, they reported cytosolic distribution of p47^{phox}, Noxo1 (homologue of p47^{phox}), Noxa1 (homologue of p67^{phox}), and p40^{phox} in these cells. Compatible with findings of Rebeleto and coworkers [40], these studies have also observed inhibition of GSIS by diphenyleneiodonium (DPI), a known inhibitor of Nox2. In another set of investigations, Oliviera and coworkers [47] studied subunit expression and functional regulatory roles of Nox2 in islet beta cell function. Using RT-PCR and/or Western blotting methods, they reported expression of various Nox2 subunits in rat pancreatic islets. Using immunohistochemical approaches, they demonstrated glucose-induced Nox2 activity in these cellular preparations. Lastly, mechanistic studies involving phorbol myristate acetate, a known activator of protein kinase C (PKC), and GF109203X, a specific inhibitor of PKC, further implicated novel roles of PKC in glucose-mediated regulation of Nox2 in pancreatic islets [47]. Together, data from the above studies have suggested acute regulatory roles for the Nox subfamily of enzymes in physiological insulin secretion. The following sections will highlight our current understanding of regulatory roles of various members of the Nox family in cytokine-mediated dysregulation of the islet beta cell, which is the primary focus of this review.

4. Regulatory Roles of Nox1 in Cytokine-Induced Dysfunction of the Beta Cell

In a series of investigations, Weaver and coworkers addressed potential roles of Nox1 in cytokine-induced islet beta-cell dysfunction. In a study published in 2012, they reported a significant increase in the expression of Nox1 in human islets, mouse islets, and clonal beta cell lines exposed to a mixture of proinflammatory cytokines [37]. A significant increase in the expression of monocyte chemoattractant protein-1 (MCP-1), ROS generation, loss in GSIS, and associated increase in cell death was also reported under these conditions. Coprovision of pharmacological inhibitors of Nox (apocynin, DPI, and a dual selective inhibitor of Nox1/Nox4) markedly attenuated cytokine-induced MCP-1 expression and ROS generation. Follow-up studies by these investigators provided additional support for regulatory roles of Nox1 in cytokine-induced dysregulation of the islet beta cell [38]. They revealed that exposure of murine beta cells to ML171, a selective inhibitor of Nox1, significantly impeded cytokine-induced ROS generation, caspase-3 activation, and cell death via apoptosis. Moreover, ML171 significantly prevented loss in GSIS induced by proinflammatory cytokines in both clonal beta cells and isolated mouse islets. Based on these data, these researchers concluded that cytokine-induced metabolic dysfunction of the islet beta cell involves Nox-1-mediated increase in ROS generation and associated intracellular oxidative stress. They also proposed that targeting of Nox-1 might serve as a valuable approach to protect proinflammatory cytokine-induced metabolic dysfunction of the beta cell in diabetes [38]. Subsequent investigations by these researchers have further affirmed the contributory roles of Nox-1 in intracellular generation of ROS and oxidative stress in pancreatic beta cells [48]. Forced expression of Nox1 in pancreatic beta cells resulted in increased generation of ROS, loss in GSIS, and increased cell death via apoptosis. It is important to note that these cellular events are comparable to those observed in beta cells under the duress of exposure to proinflammatory cytokines since shRNA-mediated suppression of Nox-1 markedly prevented deleterious effects of cytokines. Taken together, these studies have identified Nox-1 as a potential therapeutic target in the prevention of cytokine-induced islet dysfunction in diabetes.

Published evidence suggests that the 12-lipoxygenase (12-LO), which catalyzes the oxidation of fatty acids to their respective hydro peroxides, plays novel roles in cytokine-mediated dysregulation of the pancreatic islet beta cell [49,50]. In this context, studies by Weaver and coworkers demonstrated that 12-hydroxyeicosatetranoic acid (12-HETE), a product of 12-LO, significantly increased the expression of Nox-1 in human islets [37]. More importantly, NCTT-956, a selective inhibitor of 12-LO, but not its inactive analog (NCTT-695), markedly suppressed cytokine-induced Nox-1 expression, ROS generation, and caspase-3 activation in clonal beta cell preparations. Taken together, these findings provide a novel model, which implicates key regulatory roles of the 12-LO-Nox1 signaling axis in proinflammatory cytokine-induced metabolic dysregulation of the islet beta cell.

5. Regulatory Roles of Nox2 in Cytokine-Induced Dysfunction of the Beta Cell

As stated above, considerable efforts were made previously to assess the regulatory roles of phagocyte-like NADPH oxidase (Nox2) in cytokine-induced metabolic dysregulation of the islet beta cell [25,26]. Briefly, Nox2 has been shown to play key roles in phagocytosis by professional phagocytic cells (e.g., neutrophils, eosinophils, monocytes, and macrophages). The Nox2 holoenzyme is a highly regulated membrane-associated protein complex, the activation of which results in the generation of large quantities of intracellular ROS, which, in turn, promote activation of several downstream signaling events, including mitochondrial dysfunction, culminating in cell demise. Along these lines, several mechanisms have been put forth for the generation of ROS and associated oxidative stress in a variety of non-phagocytic cell types, including the islet β-cell [25,26]. As depicted in Figure 1, the Nox2 is a multicomponent system, which is comprised of membranous and cytosolic cores. The membrane-associated catalytic core is comprised of gp91^{phox} and p22^{phox}. The cytosolic core of Nox2 (also referred to as the regulatory core) is comprised of p40phox, p47^{phox}, p67^{phox}, and Rac. Following cell stimulation, the cytosolic (regulatory) core proteins translocate to the membranous compartment to associate with the catalytic core for the formation of Nox2 holoenzyme, resulting in the catalytic activation of Nox2 and generation of ROS. Several mechanisms, including phosphorylation of p40^{phox}, p47^{phox}, and p67^{phox}, have been proposed as requisite steps for the translocation of the cytosolic core to the membrane. In the case of Rac1, it appears that its activation (GTP-bound conformation), mediated by specific guanine nucleotide exchange factors (GEFs, e.g., Tiam1), favors its association with p67^{phox}, thus enabling the translocation of Rac1-p67^{phox} dimer to the membrane [11,26,51]. The experimental findings described in the following section will highlight potential contributory roles of Nox2 in the cascade of events leading to proinflammatory cytokine-induced metabolic dysfunction of the islet beta cells.

Using primary mouse islets and insulin-secreting BRIN-BD11 β-cells, Michalska and Newsholme reported significant inhibition of GSIS and an increase in the expression of p47^{phox} and iNOS in these cells following exposure to proinflammatory cytokines [52]. Interestingly, coprovision of antioxidants, such as SOD, catalase, and N-acetylcysteine (NAC), markedly suppressed effects of H_2O_2 or palmitate but not those elicited by proinflammatory cytokines. These studies also demonstrated a significant prevention of deleterious effects of cytokines and H_2O_2 following pharmacological inhibition of Nox and/or iNOS. It was concluded that H_2O_2 might play contributory roles in positive feedback redox sensitive regulation of β-cell dysfunction via its effects on Nox and iNOS [52].

Subasinghe and coworkers examined the roles of Nox2 in cytokine-induced metabolic dysfunction of the islet beta cell [53]. Specifically, they investigated the contributory roles of Rac1 in the onset of beta cell dysfunction under the duress of cytokines. They observed a significant increase in Nox2 activation, ROS generation, and in the expression of p47^{phox} subunit, but not p67^{phox} subunit, in INS-1 832/13 cells following exposure to a mixture of proinflammatory cytokines. The hypothesis that Nox2 is involved in cytokine-induced Nox2 activation and ROS generation was further confirmed using siRNA-p47^{phox}. These observations were further validated by using pharmacological inhibition of Nox2 using apocynin. Specific inhibitors of Rac1, namely NSC23766 (inhibitor of Tiam1-Rac1 signaling) and GGTI-2147 (inhibitor of prenylation of Rac1), significantly suppressed the cytokine-induced Rac1 activation and alterations in MMPT in these cells. Data accrued from these studies also suggested that the cytokine-mediated Tiam1/Rac1 signaling pathway may not be necessary for iNOS expression and NO release since NSC23766 failed to exert any significant effects on IL-1β (or a mixture of cytokines)-induced NO release in INS 832/13 cells [53]. Collectively, these findings suggested novel roles for the Tiam1-Rac1 axis in cytokine-induced ROS but not NO generation in beta cells under the duress of cytokines. In addition, these studies have provided the first evidence to suggest that prenylation of Rac1 is a requisite for cytokine-mediated effects. Based on these findings it was suggested that the combined effects of intracellularly generated NO (via activation of iNOS) and ROS (via activation of Nox2) could contribute to alterations in mitochondrial function, leading to caspase-3 activation and metabolic dysfunction of the β-cell [53].

Along these lines, Mohammed and coworkers [54] demonstrated a time-dependent phosphorylation of p47^{phox} in INS-1 832/13 cells exposed to a mixture of cytokines. A significant increase in the expression of gp91^{phox} was also noted under these conditions. Lastly, 2-Bromopalmitate, a known inhibitor of protein palmitoylation, markedly attenuated cytokine-induced, Nox2-mediated ROS generation, and iNOS-mediated NO generation [54]. Together, these studies identified palmitoyltransferase as a target for inhibition of cytomix-induced oxidative and nitrosative stress in the pancreatic beta cell. Based on the NSC23766-mediated inhibition of Rac1, it is likely that cytokine-induced activation of iNOS and Nox2 are under the control of (at least) two G proteins that require palmitoylation [53,54] (see below).

Previously published evidence suggests that phosphorylation of p47^{phox} may be mediated by PKC, a signaling event that has been implicated in the translocation of this subunit to the membranous core for Nox2 holoenzyme assembly [33,55–57]. In further support of this hypothesis, studies by Morgan et al. demonstrated partial restoration of IL-1β-induced ROS to normal levels following exposure to GF109203X, a known inhibitor of PKC [58]. These findings support the formulation for a multifactorial regulation of Nox2 subunits by proinflammatory cytokines, leading to its activation and ROS generation, and culminating in the activation of downstream signaling events involved in cell dysfunction.

Lastly, earlier studies in animal models of impaired insulin secretion and diabetes affirmed critical regulatory roles of Nox2 in cytokine-induced metabolic dysregulation of the islet. For example, Xiang and coworkers demonstrated that deficiency of Nox2 decreases beta cell destruction and preserves islet function in STZ-induced diabetes by reducing ROS production, immune response, and β-cell apoptosis [59]. Studies of Veluthakal and coworkers demonstrated that administration of NSC23766, a known inhibitor of the Tiam1-Rac1-Nox2 signaling pathway, significantly prevented the development of spontaneous diabetes in the non-obese diabetic (NOD) mice [60]. In addition, they observed that NSC23766 treatment markedly suppressed Rac1 expression, activity, and ER stress in NOD islets. Based on the findings, it was concluded that the Tiam1-Rac1-Nox2 signaling pathway plays critical regulatory roles in the onset of spontaneous diabetes in the NOD mouse model. Collectively, findings from both in vitro and in vivo provide compelling evidence for critical regulatory roles of Nox2 in proinflammatory cytokine-induced ROS generation and metabolic dysfunction culminating in the onset of islet dysfunction and diabetes.

6. Roles of Nox3, Nox4, and Nox5 in Cytokine-Induced Dysfunction of the Beta Cell

It is noteworthy that a recent literature search (Pubmed; October 2020) indicated no clear evidence of contributory roles of Nox3 in islet function. With respect to regulatory roles of Nox4, Wang et al. undertook a pharmacological approach to assess the roles of various Noxs (Nox-1, Nox-2, and Nox-4) in human pancreatic beta-cell dysfunction induced under a variety of diabetogenic conditions, including exposure to proinflammatory cytokines [41]. They demonstrated that pharmacological inhibition of Nox (using DPI, dapsone, GLX351322, and GLX481372) attenuated ROS levels, caspase activation, and loss in cell viability in human islets under the duress of glucolipotoxic conditions. ML171, a specific inhibitor of Nox1, failed to exert any significant effects on cellular dysfunction induced by diabetogenic condition, including exposure to cytokines in human islet cells. Furthermore, Phox-I2, a known inhibitor of Nox2, elicited partial protective effects induced by glucolipotoxic conditions in human islets without significantly affecting cytokine-induced dysfunction of the islet beta cell. Lastly, GLX7013114, a highly selective inhibitor of Nox4, exhibited protective effects in human beta cells under the duress of glucolipotoxicity and cytokine exposure. Based on these findings, the authors proposed that Nox4 mediates pro-apoptotic effects in intact islets under stressful conditions and that selective Nox4-inhibition may be a therapeutic strategy in type 2 diabetes [41]. Interestingly, the findings that Nox1 and Nox2 are not involved in cytokine-induced effects are in contrast to findings of the studies described in the above sections. Additional investigations are needed to explain the differences between these experimental outcomes and conclusions drawn in these studies.

It is noteworthy that, in a manner akin to Nox3, a recent review of the literature yielded very limited details on the potential regulatory roles of Nox5 in cytokine-induced dysregulation of the islet beta cell. Interestingly, however, Nox5 does appear to contribute to islet beta cell dysfunction under the duress of other pathological stimuli. For example, Bouzarki and coworkers [45] reported expression of Nox5 in somatostatin-containing delta cells in human islets under basal conditions. Selective depletion of expression of Nox5 using siRNA-Nox5 significantly attenuated GSIS, suggesting novel regulatory roles of Nox5 in physiological insulin secretion. Furthermore, long-term exposure of human islets to high glucose resulted in increased expression of Nox5 in the beta cells. Lastly, the degree of impairment in GSIS following high-fat feeding was markedly aggravated in animals in which Nox5 was conditionally increased in beta cells. Taken together, these investigations provided novel insights into the roles of Nox5 in promoting crosstalk between various cell types (i.e., the paracrine relationship between delta and beta cells) of the pancreatic islet under physiological conditions. Their findings also implicated novel roles for Nox5 in promoting vulnerability of the islet beta cell for damage under various pathological conditions [45]. As stated above, potential regulatory roles of Nox5 in proinflammatory cytokine-induced metabolic dysfunction of the islet beta cell remain to be studied further. It should also be kept in mind that Nox5 is not expressed in rats and mice [27,32]. Therefore, future studies should be focused on human islet cells in which Nox5 appears to be expressed and regulated under defined experimental conditions.

Taken together, it is evident that Nox1, Nox2, and Nox4 play critical regulatory roles in cytokine-induced metabolic dysregulation and demise of the islet beta cell. Additional investigations are needed to further explore the regulatory roles of other Nox forms, namely Nox3 and Nox5, in the cascade of events leading to islet dysfunction under the duress of cytokines. More importantly, studies in human islet cells are needed to ascertain the translational significance of these signaling pathways in the onset of beta cell dysfunction in human diabetes.

Based on the available evidence on regulatory roles of Noxs, a working model is proposed (Figure 2), which states that chronic exposure of pancreatic islet beta cells to proinflammatory cytokines results in the activation of at least three members of the Nox superfamily (Nox1, Nox2, and Nox4), leading to the generation of ROS and the onset of intracellular oxidative stress. Several lines of evidence (highlighted above) suggest that functional regulation of specific subunits of Nox is precisely mediated via post-translational modifications, including phosphorylation (e.g., p47^{phox}, p60^{phox} etc.), prenylation, and palmitoylation (e.g., Rac1). Such modifications are a requisite for their translocation to the membrane for association with the membranous core of Nox to complete holoenzyme assembly and functional activation of the Nox. As stated in the above sections, the 12-LO pathway also plays a critical regulatory role in the initiation of metabolic signals and events culminating in islet beta cell dysregulation following exposure to proinflammatory cytokines. It should be noted that Rac1 is an integral part of Nox1 and Nox2 but not Nox4 (Figure 1). Therefore, Rac1-independent mechanisms must underlie cytokine-induced regulation of Nox4. It is proposed that sustained intracellular oxidative stress and an imbalance in the ROS scavenging steps lead to mitochondrial dysfunction and activation of proapoptotic caspases (e.g., caspase-3), culminating in the cleavage and inactivation of pro-survival proteins, and resulting in accelerated beta cell dysfunction and demise [10,11].

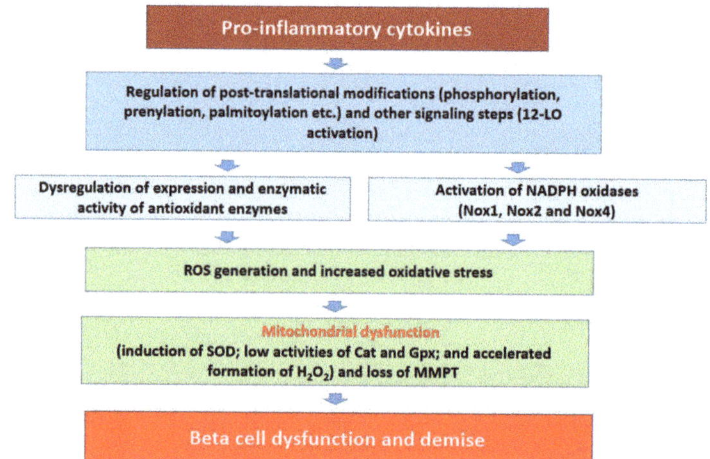

Figure 2. Potential signaling mechanisms involved in cytokine-induced Nox-mediated dysregulation of the islet beta-cell.

7. Potential Crosstalk between iNOS and Nox2 Signaling Pathways in the Onset of Cytokine-Induced Metabolic Dysregulation of the Islet Beta Cell

Peroxynitrite (PN) is generated rapidly in the cell from the interaction between NO and superoxide radicals. Increased intracellular PN levels leads to accelerated oxidation of proteins, lipids, DNA, as well as damage to intracellular organelles, including the mitochondria. Indeed, such cellular events are involved in a variety of pathological states, including cardiovascular, neuronal, and metabolic diseases [61–64]. Under healthy conditions, PN is short-lived and considered not harmful. However, under pathological conditions, a high degree of production of iNOS-derived NO and Nox-derived ROS and superoxide radicals could lead to substantially high levels of PN. As recently reviewed by Pacher and coworkers [61], a modest increase in superoxide radicals and NO by 10-fold results in a 100-fold increase in PN. The authors suggested that under proinflammatory conditions, the generation of NO and superoxide is expected to increase by 1000-fold, which results in remarkably high levels (1,000,000-fold) of PN. Indeed, such an insult would be much more damaging to the islet beta cell, which is inherently ill equipped with adequate antioxidant defense mechanisms [16,20,24,65].

Considerable debate still exists concerning potential contributory roles of PN in cytokine-induced dysfunction of the pancreatic islet beta cell. From a historical perspective of this topic, original investigations by Delaney and coworkers provided the first evidence to suggest sensitivity of human islets to PN leading to dysfunction and demise [66]. They reported that acute exposure of human islets to PN leads to inhibition of glucose oxidation and accelerated DNA damage (strand breaking). Significant alterations in cell ultrastructure, including organelle degradation, mitochondrial swelling, and matrix loss, were also noted under these conditions. Cell death analysis studies suggested necrotic, rather than apoptotic, demise of these cells following exposure to PN. Data from studies of Lakey et al. provided evidence for critical regulatory roles of PN in cytokine-mediated dysfunction of human pancreatic islets [67]. They reported high levels of nitrotyrosine, a marker of PN, in islet cells exposed to a mixture of proinflammatory cytokines. In a manner akin to the effects of cytokines, provision of exogenous PN led to nitrotyrosine formation and cell dysfunction. Lastly, co-provision of guanidinoethyldisulphide (GED), a known inhibitor of iNOS and scavenger of PN, attenuated cytokine-induced NO release, H_2O_2 production, nitrotyrosine formation, and associated cell dysfunction. Based on these observations, it was concluded that PN formation is causal to cytokine-mediated effects on human islets. Subsequent studies by Suarez-Pinzon et al. demonstrated that GED significantly reduced the onset of diabetes in the NOD mouse model [68]. Furthermore, GED markedly suppressed NO and nitrotyrosine formation and cell demise in NOD mouse islets incubated with proinflammatory cytokines. It was concluded that increases in intracellular superoxide radicals and NO levels culminate in the formation of PN, which, in turn, leads to beta cell destruction in autoimmune diabetes. Further investigations by Mabley and coworkers [69] affirmed the therapeutic efficacy of GED in preventing the onset of islet dysfunction in these experimental models. Contrary to the evidence described above, more recent findings from Corbett's laboratory have suggested that NO, but not PN, mediates cytokine-induced dysfunction of the islet. Evidence is also presented by these researchers that endogenous cytosolic peroxiredoxin 1 (Prdx1) affords protection to the beta cell against intracellularly generated ROS and reactive nitrogen species (RNS) [70]. Collectively, it is evident from the above narrative that cytokine-induced NO and ROS levels exert deleterious effects singly, or in combination, on beta cell function. As stated above, it is important to note that data accrued from investigations involving human islets revealed that the human beta cells are not equipped to generate NO as they do not express iNOS. However, they have been shown to elicit sensitivity to RNS, when co-provided exogenously, resulting in increased interaction with super oxide radicals, intracellularly culminating in beta cell dysfunction. [17,21–23]. These aspects must be kept in mind in the interpretation of experimental data from earlier investigations, and in planning future studies to further decipher roles of this signaling module and PN in the cascade of events leading to cytokine-induced damage to the islet beta cell.

8. Restoration of Intracellular Redox Environment Prevents Cytokine-Induced Metabolic Defects in the Beta Cell

Despite the rapid advances in the field, several knowledge gaps still exist, specifically in addressing the fact that the beta cell antioxidant capacity and its ability to scavenge H_2O_2 are relatively low compared to other cell types. Extant studies have utilized many approaches, including provision of antioxidants to "rescue" the beta cell against noxious effects of diabetogenic stimuli, including proinflammatory cytokines. For example, using insulin-secreting BRIN-BD11 beta cells, Michalska et al. reported beneficial effects of antioxidants, such as SOD, catalase, and NAC, against deleterious effects of H_2O_2 but not cytokines [52]. In a series of methodical investigations Tran et al. demonstrated that adenoviral overexpression of glutamylcysteine ligase, an enzyme involved in the de novo biosynthesis of glutathione, protects pancreatic islets against IL-1β-induced loss in GSIS; such effects were shown to be via an increase in intracellular reduced glutathione (GSH) levels [71]. Studies by Gurgul and coworkers provided evidence for significant protection of insulin-secreting RINm5F cells overexpressing mitochondrial catalase against proinflammatory cytokine-induced cell death; these findings affirm contributory roles for mitochondrial ROS in cytokine-mediated effects [72]. Using stable expression and suppression of MnSOD in RINm5F cells, Lortz and associates further validated the hypothesis that an imbalance between superoxide generation and H_2O_2 detoxification enzymes dictates the vulnerability of beta cells to cytokine-induced damage [73]. Overexpression of catalase in the mitochondria has also been shown to afford protection against cytokine-induced nitro-oxidative stress and demise in insulin-producing RINm5F cells [19]. Interestingly, expression of an endoplasmic reticulum-targeted and luminal-active catalase variant (ER-catalase N244) provided protection in INS-1E cells against H_2O_2- but not cytokine-induced toxicity [74]. Lastly, studies of Mehmeti and coworkers [75] revealed that overexpression of mitochondrial-specific catalase (MitoCatalase) prevented cytokine-induced alterations in Bax/Bcl-2, and the downstream signaling events, including cytochrome C release and activation of executioner caspases 3 and 9. Indeed, data from the above investigations provide support for the overall hypothesis that deficiencies and/or alterations in ROS scavenging mechanisms within the mitochondrial compartment increase the susceptibility of the islet beta cell to cytokine-induced damage.

Several other candidate genes were examined for their protective effects of beta cells against cytokine insult. For example, Stancill and coworkers recently reported that pharmacological inhibition of Prdx1 (by conoidin A) or siRNA-mediated depletion of expression of endogenous Prdx1 significantly increased the vulnerability of clonal beta cells and rat islets to H_2O_2 and PN. Based on these findings, the authors concluded that Prdx1 provides defense against intracellularly generated nitroso and oxidative stress in the cytokine-challenged beta cell [70]. In this context, Wolf and coworkers investigated the potential cytoprotective roles of mitochondrial Prix III, a thioredoxin-dependent peroxide reductase, against oxidative and nitrosative stress in rat insulinoma cells either over- or under-expressing Prdx III, under the duress of a variety of stimuli, including H_2O_2 and proinflammatory cytokines [76]. Their findings revealed a significant resistance in Prdx III-expressing cells to these stimuli as evidenced by a marked suppression of iNOS gene expression and the downstream signaling events, including caspase-9 and caspase-3 activation. Together, the above investigations revealed key protective roles of the Prdx family of proteins against cytokine-mediated dysregulation of the islet beta cell. Lastly, using knockout animal models, studies from Lammert's laboratory demonstrated that antioxidant protein DJ-1, which is encoded by the Parkinson's disease gene PARK7, affords protection of beta cells against proinflammatory cytokine-mediated cell dysfunction and demise [77,78]. Altogether, thee examples of studies cited above affirm support for the proposal that maintenance (or boosting) of intracellular antioxidant capacity (defense mechanisms) of the beta cell represents a viable therapeutic option to protect the beta cell against cytokine insult.

9. Conclusions

A growing body of evidence, in in vitro and in vivo model systems, implicates intracellularly generated oxidative stress as one of the contributing factors in the cascade of events leading to cytokine-induced metabolic dysfunction and demise of the islet beta cell. In this review, I attempted to summarize the known evidence in support of regulatory roles of Noxs, specifically Nox1 and Nox2, in cytokine-induced alterations in the metabolic functions of the islet. Evidence in favor of a regulatory role for Nox4 is emerging, but potential roles of Nox3 and Nox5 in this signaling cascade remain relatively poorly understood. Several pharmacological approaches have been employed to decipher the roles of Noxs, iNOS, and 12-LO as mediators of cytokine-induced metabolic dysregulation of the islet beta cell (Table 1). Indeed, such approaches have provided useful insights into these pathways as potential targets to prevent/halt the metabolic defects. Potential regulatory roles of PN as a mediator of beta cell damage under the duress of cytokines remains to be examined further. In addition, much is unknown with regard to signaling mechanisms and regulatory proteins/factors that promote crosstalk between iNOS-12-LO-Nox2 signaling pathways in the cascade of events leading to cytokine dysregulation of the islet beta cell. These aspects need to be addressed further.

Table 1. Inhibitors of the Nox, iNOS, and 12-LO signaling pathways employed in studies highlighted in this review.

Inhibitor	Mechanism(s) of Action	Reference
GGTI-2147	Inhibitor of prenylation of G proteins (Rac1)	[53]
NSC23766	Inhibitor of Tiam1-Rac1 signaling pathway	[53]
Manumycin	Pan inhibitor of farnesylation of G proteins (Ras)	[79–81]
Damnacanthal	Pan inhibitor of farnesylation of G proteins (Ras)	[80,81]
Cerulenin	Inhibitor of palmitoylation of G proteins	[79]
2-bromopalmitate	Inhibitor of palmitoylation of G proteins	[54,79]
DPI	Pan inhibitor of Noxs	[37,41,58]
Apocynin	Pan inhibitor of Noxs2	[37,53,54]
Gp91ds-tat	Inhibitor of Nox2	[82]
Guanidinoethylsulphide	Inhibitor of iNOS and scavenger of PN	[67,68]
GLX7013114	Specific inhibitor of Nox4	[41]
GLX351322	Selective inhibitor of Nox4 (inhibits other Noxs)	[41]
GLX481372	Selective inhibitor of Nox4 (inhibits other Noxs)	[41]
Dapsone	Inhibits expression/activity of Nox4 and DUOX1	[41]
ML171	Inhibitor of Nox1	[38,41]
Phox-I2	Inhibitor of Nox2	[41]
ML351	Inhibitor of 12/15-LO	[83]
NCTT-956	Inhibitor of 12-LO	[37,84]
GF109203X	Inhibitor of PKC	[58]

Based on the evidence, principally accrued in our laboratory on the regulatory roles of small G proteins in the generation of iNOS-derived ROS and Nox2-derived ROS, I propose a model for a potential crosstalk between these signaling pathways in eliciting damage to the rat islet beta cell following exposure to proinflammatory cytokines (Figure 3). Briefly, using pharmacological approaches, it was demonstrated that IL-1β-induced iNOS expression and NO release are under the control of H-Ras, a small G protein belonging to the Ras superfamily of G proteins [80,81]. These findings were further confirmed with bacterial toxins that promote glucosylation and inactivation of small G proteins [80,81]. Interestingly, data accrued from the investigations involving bacterial toxins suggested that activation of Rho G proteins (e.g., Rac1) is not a requisite for IL-1β-induced iNOS expression and NO acid release [81]. Subsequent studies involving specific inhibitors of Rac1 (e.g., NSC23766) further affirmed the postulation that Rac1 is involved in ROS generation but not NO production in a cytokine-challenged beta cell [53]. The combined effects of intracellularly generated NO (due to activation of iNOS) and ROS (due activation of Nox) could contribute to maximal damage of the mitochondrial membrane properties leading to metabolic dysfunction of the beta cell (Figures 2 and 3). It is emphasized that this model, involving H-Ras, may not be applicable to the human beta cells since they do not express the iNOS-NO signaling pathway, and yet exert sensitivity to NO, which is secreted from activated immune cells. The overall concept of intracellular generation of PN and its effects on mitochondrial dysregulation, loss in GSIS, and apoptotic demise of the islet beta cell under the duress of cytokines was proposed earlier [85]. Additional studies are needed to further substantiate this formulation. It is my hope that the future advances in the field of Nox biology will help not only in our current understanding of this class of enzymes but also in the development of small-molecule compounds (via combinatorial chemistry approaches) with a high degree of specificity to inhibit the iNOS-Nox-LO pathways. The development of methodologies to boost the overall antioxidant capacity of the islet beta cell is also warranted. Lastly, novel approaches to accelerate detoxification of intracellularly generated H_2O_2 in specific subcellular compartments (e.g., cytosol and mitochondria) might prove valuable in preventing cytokine-mediated dysfunction and demise of the pancreatic islet beta cell.

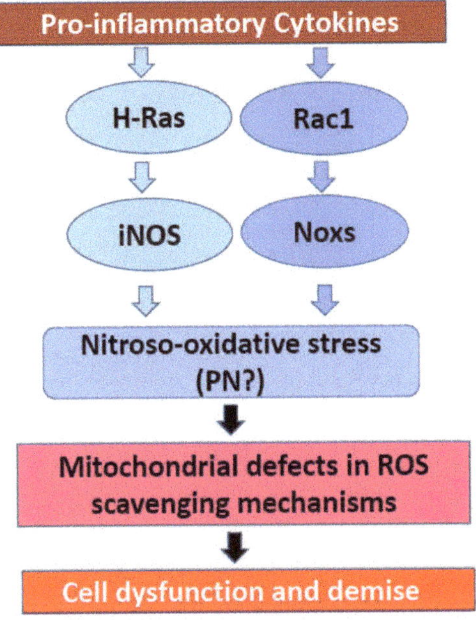

Figure 3. Potential involvement of small G proteins in cytokine-induced NO release and ROS formation leading to mitochondrial dysfunction and demise of the rat islet beta cell.

Funding: This research is supported by a Merit Review (I01 BX004663) and a Senior Research Career Scientist (IK6 BX005383) awards from the US Department of Veterans Affairs, and a grant from the National Eye Institute (EY022230). The author also received support from the Juvenile Diabetes Foundation International.

Acknowledgments: The author thanks the former and current members of his research group and collaborators for their contribution to the work highlighted herein.

Conflicts of Interest: The author declares no conflict of interest.

Abbreviations

DPI	diphenyleneiodonium
Duox1	dual oxidase1
Duox2	dual oxidase2
EF hands	calcium binding domains or motifs
ER stress	endoplasmic reticulum stress
FTase	farnesyltransferase
GEFs	guanine nucleotide exchange factors
GSH	reduced glutathione
GSIS	glucose-stimulated insulin secretion
12-HETE	12-hydroxyeicosatetranoic acid
IFNγ	interferon γ
IL-1β	interleukin 1β
iNOS	inducible nitric oxide synthase
12-LO	12-lipoxygenase
MCP-1	monocyte chemoattractant protein-1
MMPT	mitochondrial membrane pore transition
Nrf2	nuclear factor erythroid 2-related factor 2
NO	nitric oxide
Nox1	NADPH oxidase 1
Nox2	NADPH oxidase 2
Nox3	NADPH oxidase 3
Nox4	NADPH oxidase 4
Nox5	NADPH oxidase 5
PKC	protein kinase C
PN	peroxynitrite
Prdx1	peroxyredoxin1
PrdxIII	peroxyredoxin III
Prdx6	peroxyredoxin 6
siRNA	small interfering RNA
SOD	superoxide dismutase
TNFα	tumor necrosis factorα
NAC	N-acetylcysteine
NSC23766	N6-[2-[4-(Diethylamino)-1-methylbu-tyl]amino]-6-methyl-4-pyrimidinyl]-2-methyl-4,6-qu-inolinediamine trihydrochloride
NOD	non-obese diabetic
Rac1	Ras-related C3 botulinum toxin substrate 1
RNS	reactive nitrogen species
ROS	reactive oxygen species
T1DM	type 1 diabetes mellitus
Tiam1	T-cell lymphoma invasion and metastasis-inducing protein 1

References

1. Nerup, J.; Mandrup-Poulsen, T.; Mølvig, J.; Helqvist, S.; Wogensen, L.; Egeberg, J. Mechanisms of pancreatic beta-cell destruction in type I diabetes. *Diabetes Care* **1988**, *11* (Suppl. 1), 16–23.
2. Kaminitz, A.; Stein, J.; Yaniv, I.; Askenasy, N. The vicious cycle of apoptotic beta-cell death in type 1 diabetes. *Immunol. Cell Biol.* **2007**, *85*, 582–589. [CrossRef]
3. Mehmeti, I.; Gurgul-Convey, E.; Lenzen, S.; Lortz, S. Induction of the intrinsic apoptosis pathway in insulin-secreting cells is dependent on oxidative damage of mitochondria but independent of caspase-12 activation. *Biochim. Biophys. Acta* **2011**, *1813*, 1827–1835. [CrossRef]
4. Zhong, L.; Tran, T.; Baguley, T.D.; Lee, S.J.; Henke, A.; To, A.; Li, S.; Yu, S.; Grieco, F.A.; Roland, J.; et al. A novel inhibitor of inducible NOS dimerization protects against cytokine-induced rat beta cell dysfunction. *Br. J. Pharmacol.* **2018**, *175*, 3470–3485. [CrossRef]
5. Berchtold, L.A.; Prause, M.; Størling, J.; Mandrup-Poulsen, T. Cytokines and pancreatic β-Cell apoptosis. *Adv. Clin. Chem.* **2016**, *75*, 99–158. [CrossRef]
6. Brun, T.; Maechler, P. Beta-cell mitochondrial carriers and the diabetogenic stress response. *Biochim. Biophys. Acta* **2016**, *1863*, 2540–2549. [CrossRef]
7. Eizirik, D.L.; Colli, M.L.; Ortis, F. The role of inflammation in insulitis and beta-cell loss in type 1 diabetes. *Nat. Rev. Endocrinol.* **2009**, *5*, 219–226. [CrossRef]
8. Rabinovitch, A.; Suarez-Pinzon, W.L. Cytokines and their roles in pancreatic islet beta-cell destruction and insulin-dependent diabetes mellitus. *Biochem. Pharmacol.* **1998**, *55*, 1139–1149. [CrossRef]
9. Padgett, L.E.; Broniowska, K.A.; Hansen, P.A.; Corbett, J.A.; Tse, H.M. The role of reactive oxygen species and proinflammatory cytokines in type 1 diabetes pathogenesis. *Ann. N. Y. Acad. Sci.* **2013**, *1281*, 16–35. [CrossRef]
10. Veluthakal, R.; Arora, D.K.; Goalstone, M.L.; Kowluru, R.A.; Kowluru, A. Metabolic stress induces Caspase-3 mediated degradation and inactivation of farnesyl and geranylgeranyl transferase activities in pancreatic β-Cells. *Cell. Physiol. Biochem.* **2016**, *39*, 2110–2120. [CrossRef]
11. Kowluru, A. Role of G-proteins in islet function in health and diabetes. *Diabetes Obes. Metab.* **2017**, *19* (Suppl. 1), 63–75. [CrossRef] [PubMed]
12. Eguchi, K.; Manabe, I. Macrophages and islet inflammation in type 2 diabetes. *Diabetes Obes. Metab.* **2013**, *15* (Suppl. 3), 152–158. [CrossRef] [PubMed]
13. Hasnain, S.Z.; Prins, J.B.; McGuckin, M.A. Oxidative and endoplasmic reticulum stress in β-cell dysfunction in diabetes. *J. Mol. Endocrinol.* **2016**, *56*, R33–R54. [CrossRef]
14. Ogihara, T.; Mirmira, R.G. An islet in distress: β cell failure in type 2 diabetes. *J. Diabetes Investig.* **2010**, *1*, 123–133. [CrossRef]
15. Cardozo, A.K.; Ortis, F.; Storling, J.; Feng, Y.M.; Rasschaert, J.; Tonnesen, M.; van Eylen, F.; Mandrup-Poulsen, T.; Herchuelz, A.; Eizirik, D.L. Cytokines downregulate the sarcoendoplasmic reticulum pump Ca^{2+} ATPase 2b and deplete endoplasmic reticulum Ca^{2+}, leading to induction of endoplasmic reticulum stress in pancreatic beta-cells. *Diabetes* **2005**, *54*, 452–461. [CrossRef]
16. Lenzen, S.; Drinkgern, J.; Tiedge, M. Low antioxidant enzyme gene expression in pancreatic islets compared with various other mouse tissues. *Free Radic. Biol. Med.* **1996**, *20*, 463–466. [CrossRef]
17. Gurgul-Convey, E.; Mehmeti, I.; Plötz, T.; Jörns, A.; Lenzen, S. Sensitivity profile of the human EndoC-βH1 beta cell line to proinflammatory cytokines. *Diabetologia* **2016**, *59*, 2125–2133. [CrossRef]
18. Miki, A.; Ricordi, C.; Sakuma, Y.; Yamamoto, T.; Misawa, R.; Mita, A.; Molano, R.D.; Vaziri, N.D.; Pileggi, A.; Ichii, H. Divergent antioxidant capacity of human islet cell subsets: A potential cause of beta-cell vulnerability in diabetes and islet transplantation. *PLoS ONE* **2018**, *13*, e0196570. [CrossRef]
19. Gurgul-Convey, E.; Mehmeti, I.; Lortz, S.; Lenzen, S. Cytokine toxicity in insulin-producing cells is mediated by nitro-oxidative stress-induced hydroxyl radical formation in mitochondria. *J. Mol. Med.* **2011**, *89*, 785–798. [CrossRef]
20. Lenzen, S. Chemistry and biology of reactive species with special reference to the antioxidative defence status in pancreatic β-cells. *Biochim. Biophys. Acta Gen. Subj.* **2017**, *1861*, 1929–1942. [CrossRef]
21. Brozzi, F.; Nardelli, T.R.; Lopes, M.; Millard, I.; Barthson, J.; Igoillo-Esteve, M.; Grieco, F.A.; Villate, O.; Oliveira, J.M.; Casimir, M.; et al. Cytokines induce endoplasmic reticulum stress in human, rat and mouse beta cells via different mechanisms. *Diabetologia* **2015**, *58*, 2307–2316. [CrossRef] [PubMed]

22. Eizirik, D.L.; Sandler, S.; Welsh, N.; Cetkovic-Cvrlje, M.; Nieman, A.; Geller, D.A.; Pipeleers, D.G.; Bendtzen, K.; Hellerström, C. Cytokines suppress human islet function irrespective of their effects on nitric oxide generation. *J. Clin. Investig.* **1994**, *93*, 1968–1974. [CrossRef]

23. Welsh, N.; Eizirik, D.L.; Sandler, S. Nitric oxide and pancreatic beta-cell destruction in insulin dependent diabetes mellitus: Don't take NO for an answer. *Autoimmunity* **1994**, *18*, 285–290. [CrossRef] [PubMed]

24. Acharya, J.D.; Ghaskadbi, S.S. Islets and their antioxidant defense. *Islets* **2010**, *2*, 225–235. [CrossRef]

25. Newsholme, P.; Morgan, D.; Rebelato, E.; Oliveira-Emilio, H.C.; Procopio, J.; Curi, R.; Carpinelli, A. Insights into the critical role of NADPH oxidase(s) in the normal and dysregulated pancreatic beta cell. *Diabetologia* **2009**, *52*, 2489–2498. [CrossRef]

26. Kowluru, A.; Kowluru, R.A. Phagocyte-like NADPH oxidase [Nox2] in cellular dysfunction in models of glucolipotoxicity and diabetes. *Biochem. Pharmacol.* **2014**, *88*, 275–283. [CrossRef]

27. Buvelot, H.; Jaquet, V.; Krause, K.H. Mammalian NADPH Oxidases. *Methods Mol. Biol.* **2019**, *1982*, 17–36. [CrossRef]

28. Laddha, A.P.; Kulkarni, Y.A. NADPH oxidase: A membrane-bound enzyme and its inhibitors in diabetic complications. *Eur. J. Pharmacol.* **2020**, *881*, 173206. [CrossRef]

29. Schröder, K. NADPH oxidases: Current aspects and tools. *Redox Biol.* **2020**, *34*, 101512. [CrossRef]

30. Guichard, C.; Moreau, R.; Pessayre, D.; Epperson, T.K.; Krause, K.H. NOX family NADPH oxidases in liver and in pancreatic islets: A role in the metabolic syndrome and diabetes? *Biochem. Soc. Trans.* **2008**, *36*, 920–929. [CrossRef]

31. Grandvaux, N.; Soucy-Faulkner, A.; Fink, K. Innate host defense: Nox and Duox on phox's tail. *Biochimie* **2007**, *89*, 1113–1122. [CrossRef]

32. Bedard, K.; Jaquet, V.; Krause, K.H. NOX5: From basic biology to signaling and disease. *Free Radic. Biol. Med.* **2012**, *52*, 725–734. [CrossRef]

33. Leto, T.L.; Morand, S.; Hurt, D.; Ueyama, T. Targeting and regulation of reactive oxygen species generation by Nox family NADPH oxidases. *Antioxid. Redox Signal.* **2009**, *11*, 2607–2619. [CrossRef]

34. Kato, K.; Hecker, L. NADPH oxidases: Pathophysiology and therapeutic potential in age-associated pulmonary fibrosis. *Redox Biol.* **2020**, *33*, 101541. [CrossRef]

35. Vignais, P.V. The superoxide-generating NADPH oxidase: Structural aspects and activation mechanism. *Cell. Mol. Life Sci.* **2002**, *59*, 1428–1459. [CrossRef]

36. Leusen, J.H.; Verhoeven, A.J.; Roos, D. Interactions between the components of the human NADPH oxidase: A review about the intrigues in the phox family. *Front. Biosci.* **1996**, *1*, d72–d90. [CrossRef]

37. Weaver, J.R.; Holman, T.R.; Imai, Y.; Jadhav, A.; Kenyon, V.; Maloney, D.J.; Nadler, J.L.; Rai, G.; Simeonov, A.; Taylor-Fishwick, D.A. Integration of pro-inflammatory cytokines, 12-lipoxygenase and NOX-1 in pancreatic islet beta cell dysfunction. *Mol. Cell. Endocrinol.* **2012**, *358*, 88–95. [CrossRef]

38. Weaver, J.R.; Grzesik, W.; Taylor-Fishwick, D.A. Inhibition of NADPH oxidase-1 preserves beta cell function. *Diabetologia* **2015**, *58*, 113–121. [CrossRef]

39. Imai, Y.; Dobrian, A.D.; Weaver, J.R.; Butcher, M.J.; Cole, B.K.; Galkina, E.V.; Morris, M.A.; Taylor-Fishwick, D.A.; Nadler, J.L. Interaction between cytokines and inflammatory cells in islet dysfunction, insulin resistance and vascular disease. *Diabetes Obes. Metab.* **2013**, *15* (Suppl. 3), 117–129. [CrossRef] [PubMed]

40. Rebelato, E.; Mares-Guia, T.R.; Graciano, M.F.; Labriola, L.; Britto, L.R.; Garay-Malpartida, H.M.; Curi, R.; Sogayar, M.C.; Carpinelli, A.R. Expression of NADPH oxidase in human pancreatic islets. *Life Sci.* **2012**, *91*, 244–249. [CrossRef]

41. Wang, X.; Elksnis, A.; Wikström, P.; Walum, E.; Welsh, N.; Carlsson, P.O. The novel NADPH oxidase 4 selective inhibitor GLX7013114 counteracts human islet cell death in vitro. *PLoS ONE* **2018**, *13*, e0204271. [CrossRef] [PubMed]

42. Syed, I.; Kyathanahalli, C.N.; Jayaram, B.; Govind, S.; Rhodes, C.J.; Kowluru, R.A.; Kowluru, A. Increased phagocyte-like NADPH oxidase and ROS generation in type 2 diabetic ZDF rat and human islets: Role of Rac1-JNK1/2 signaling pathway in mitochondrial dysregulation in the diabetic islet. *Diabetes* **2011**, *60*, 2843–2852. [CrossRef] [PubMed]

43. Li, N.; Li, B.; Brun, T.; Deffert-Delbouille, C.; Mahiout, Z.; Daali, Y.; Ma, X.J.; Krause, K.H.; Maechler, P. NADPH oxidase NOX2 defines a new antagonistic role for reactive oxygen species and cAMP/PKA in the regulation of insulin secretion. *Diabetes* **2012**, *61*, 2842–2850. [CrossRef] [PubMed]

44. Anvari, E.; Wikström, P.; Walum, E.; Welsh, N. The novel NADPH oxidase 4 inhibitor GLX351322 counteracts glucose intolerance in high-fat diet-treated C57BL/6 mice. *Free Radic. Res.* **2015**, *49*, 1308–1318. [CrossRef]

45. Bouzakri, K.; Veyrat-Durebex, C.; Holterman, C.; Arous, C.; Barbieux, C.; Bosco, D.; Altirriba, J.; Alibashe, M.; Tournier, B.B.; Gunton, J.E.; et al. Beta-cell-specific expression of nicotinamide adenine dinucleotide phosphate oxidase 5 aggravates high-fat diet-induced impairment of islet insulin secretion in mice. *Antioxid. Redox Signal.* **2020**, *32*, 618–635. [CrossRef]

46. Uchizono, Y.; Takeya, R.; Iwase, M.; Sasaki, N.; Oku, M.; Imoto, H.; Iida, M.; Sumimoto, H. Expression of isoforms of NADPH oxidase components in rat pancreatic islets. *Life Sci.* **2006**, *80*, 133–139. [CrossRef]

47. Oliveira, H.R.; Verlengia, R.; Carvalho, C.R.; Britto, L.R.; Curi, R.; Carpinelli, A.R. Pancreatic beta-cells express phagocyte-like NAD(P)H oxidase. *Diabetes* **2003**, *52*, 1457–1463. [CrossRef]

48. Weaver, J.; Taylor-Fishwick, D.A. Relationship of NADPH Oxidase-1 expression to beta cell dysfunction induced by inflammatory cytokines. *Biochem. Biophys. Res. Commun.* **2017**, *485*, 290–294. [CrossRef]

49. Bleich, D.; Chen, S.; Gu, J.L.; Nadler, J.L. The role of 12-lipoxygenase in pancreatic -cells (Review). *Int. J. Mol. Med.* **1998**, *1*, 265–272. [CrossRef]

50. Prasad, K.M.; Thimmalapura, P.R.; Woode, E.A.; Nadler, J.L. Evidence that increased 12-lipoxygenase expression impairs pancreatic beta cell function and viability. *Biochem. Biophys. Res. Commun.* **2003**, *308*, 427–432. [CrossRef]

51. Kowluru, A. GPCRs, G Proteins, and their impact on β-cell function. *Compr. Physiol.* **2020**, *10*, 453–490. [CrossRef] [PubMed]

52. Michalska, M.; Wolf, G.; Walther, R.; Newsholme, P. Effects of pharmacological inhibition of NADPH oxidase or iNOS on pro-inflammatory cytokine, palmitic acid or H2O2-induced mouse islet or clonal pancreatic β-cell dysfunction. *Biosci. Rep.* **2010**, *30*, 445–453. [CrossRef] [PubMed]

53. Subasinghe, W.; Syed, I.; Kowluru, A. Phagocyte-like NADPH oxidase promotes cytokine-induced mitochondrial dysfunction in pancreatic β-cells: Evidence for regulation by Rac1. *Am. J. Physiol. Regul. Integr. Comp. Physiol.* **2011**, *300*, R12–R20. [CrossRef] [PubMed]

54. Mohammed, A.M.; Syeda, K.; Hadden, T.; Kowluru, A. Upregulation of phagocyte-like NADPH oxidase by cytokines in pancreatic beta-cells: Attenuation of oxidative and nitrosative stress by 2-bromopalmitate. *Biochem. Pharmacol.* **2013**, *85*, 109–114. [CrossRef] [PubMed]

55. Belambri, S.A.; Rolas, L.; Raad, H.; Hurtado-Nedelec, M.; Dang, P.M.; El-Benna, J. NADPH oxidase activation in neutrophils: Role of the phosphorylation of its subunits. *Eur. J. Clin. Investig.* **2018**, *48* (Suppl. 2), e12951. [CrossRef] [PubMed]

56. El-Benna, J.; Dang, P.M.; Gougerot-Pocidalo, M.A.; Marie, J.C.; Braut-Boucher, F. p47phox, the phagocyte NADPH oxidase/NOX2 organizer: Structure, phosphorylation and implication in diseases. *Exp. Mol. Med.* **2009**, *41*, 217–225. [CrossRef]

57. Rastogi, R.; Geng, X.; Li, F.; Ding, Y. NOX activation by subunit interaction and underlying mechanisms in disease. *Front. Cell. Neurosci.* **2016**, *10*, 301. [CrossRef]

58. Morgan, D.; Oliveira-Emilio, H.R.; Keane, D.; Hirata, A.E.; Santos da Rocha, M.; Bordin, S.; Curi, R.; Newsholme, P.; Carpinelli, A.R. Glucose, palmitate and pro-inflammatory cytokines modulate production and activity of a phagocyte-like NADPH oxidase in rat pancreatic islets and a clonal beta cell line. *Diabetologia* **2007**, *50*, 359–369. [CrossRef]

59. Xiang, F.L.; Lu, X.; Strutt, B.; Hill, D.J.; Feng, Q. NOX2 deficiency protects against streptozotocin-induced beta-cell destruction and development of diabetes in mice. *Diabetes* **2010**, *59*, 2603–2611. [CrossRef]

60. Veluthakal, R.; Sidarala, V.; Kowluru, A. NSC23766, a known inhibitor of Tiam1-Rac1 signaling module, prevents the onset of type 1 diabetes in the NOD mouse model. *Cell. Physiol. Biochem.* **2016**, *39*, 760–767. [CrossRef]

61. Pacher, P.; Beckman, J.S.; Liaudet, L. Nitric oxide and peroxynitrite in health and disease. *Physiol. Rev.* **2007**, *87*, 315–424. [CrossRef]

62. Ghasemi, M.; Mayasi, Y.; Hannoun, A.; Eslami, S.M.; Carandang, R. Nitric oxide and mitochondrial function in neurological diseases. *Neuroscience* **2018**, *376*, 48–71. [CrossRef] [PubMed]

63. Islam, B.U.; Habib, S.; Ahmad, P.; Allarakha, S.; Moinuddin; Ali, A. Pathophysiological role of peroxynitrite induced DNA damage in human diseases: A special focus on poly(ADP-ribose) polymerase (PARP). *Indian J. Clin. Biochem.* **2015**, *30*, 368–385. [CrossRef] [PubMed]

64. Kanaan, G.N.; Harper, M.E. Cellular redox dysfunction in the development of cardiovascular diseases. *Biochim. Biophys. Acta Gen. Subj.* **2017**, *1861*, 2822–2829. [CrossRef]

65. Poitout, V.; Robertson, R.P. Glucolipotoxicity: Fuel excess and beta-cell dysfunction. *Endocr. Rev.* **2008**, *29*, 351–366. [CrossRef]

66. Delaney, C.A.; Tyrberg, B.; Bouwens, L.; Vaghef, H.; Hellman, B.; Eizirik, D.L. Sensitivity of human pancreatic islets to peroxynitrite-induced cell dysfunction and death. *FEBS Lett.* **1996**, *394*, 300–306. [CrossRef]

67. Lakey, J.R.; Suarez-Pinzon, W.L.; Strynadka, K.; Korbutt, G.S.; Rajotte, R.V.; Mabley, J.G.; Szabó, C.; Rabinovitch, A. Peroxynitrite is a mediator of cytokine-induced destruction of human pancreatic islet beta cells. *Lab. Investig.* **2001**, *81*, 1683–1692. [CrossRef]

68. Suarez-Pinzon, W.L.; Mabley, J.G.; Strynadka, K.; Power, R.F.; Szabó, C.; Rabinovitch, A. An inhibitor of inducible nitric oxide synthase and scavenger of peroxynitrite prevents diabetes development in NOD mice. *J. Autoimmun.* **2001**, *16*, 449–455. [CrossRef]

69. Mabley, J.G.; Southan, G.J.; Salzman, A.L.; Szabó, C. The combined inducible nitric oxide synthase inhibitor and free radical scavenger guanidinoethyldisulfide prevents multiple low-dose streptozotocin-induced diabetes in vivo and interleukin-1beta-induced suppression of islet insulin secretion in vitro. *Pancreas* **2004**, *28*, E39–E44. [CrossRef]

70. Stancill, J.S.; Happ, J.T.; Broniowska, K.A.; Hogg, N.; Corbett, J.A. Peroxiredoxin 1 plays a primary role in protecting pancreatic β-cells from hydrogen peroxide and peroxynitrite. *Am. J. Physiol. Regul. Integr. Comp. Physiol.* **2020**, *318*, R1004–R1013. [CrossRef]

71. Tran, P.O.; Parker, S.M.; LeRoy, E.; Franklin, C.C.; Kavanagh, T.J.; Zhang, T.; Zhou, H.; Vliet, P.; Oseid, E.; Harmon, J.S.; et al. Adenoviral overexpression of the glutamylcysteine ligase catalytic subunit protects pancreatic islets against oxidative stress. *J. Biol. Chem.* **2004**, *279*, 53988–53993. [CrossRef] [PubMed]

72. Gurgul, E.; Lortz, S.; Tiedge, M.; Jörns, A.; Lenzen, S. Mitochondrial catalase overexpression protects insulin-producing cells against toxicity of reactive oxygen species and proinflammatory cytokines. *Diabetes* **2004**, *53*, 2271–2280. [CrossRef]

73. Lortz, S.; Gurgul-Convey, E.; Lenzen, S.; Tiedge, M. Importance of mitochondrial superoxide dismutase expression in insulin-producing cells for the toxicity of reactive oxygen species and proinflammatory cytokines. *Diabetologia* **2005**, *48*, 1541–1548. [CrossRef] [PubMed]

74. Lortz, S.; Lenzen, S.; Mehmeti, I. Impact of scavenging hydrogen peroxide in the endoplasmic reticulum for β cell function. *J. Mol. Endocrinol.* **2015**, *55*, 21–29. [CrossRef] [PubMed]

75. Mehmeti, I.; Lenzen, S.; Lortz, S. Modulation of Bcl-2-related protein expression in pancreatic beta cells by pro-inflammatory cytokines and its dependence on the antioxidative defense status. *Mol. Cell. Endocrinol.* **2011**, *332*, 88–96. [CrossRef]

76. Wolf, G.; Aumann, N.; Michalska, M.; Bast, A.; Sonnemann, J.; Beck, J.F.; Lendeckel, U.; Newsholme, P.; Walther, R. Peroxiredoxin III protects pancreatic ß cells from apoptosis. *J. Endocrinol.* **2010**, *207*, 163–175. [CrossRef] [PubMed]

77. Jain, D.; Weber, G.; Eberhard, D.; Mehana, A.E.; Eglinger, J.; Welters, A.; Bartosinska, B.; Jeruschke, K.; Weiss, J.; Päth, G.; et al. DJ-1 protects pancreatic beta cells from cytokine- and streptozotocin-mediated cell death. *PLoS ONE* **2015**, *10*, e0138535. [CrossRef]

78. Eberhard, D.; Lammert, E. The role of the antioxidant protein DJ-1 in type 2 diabetes mellitus. *Adv. Exp. Med. Biol.* **2017**, *1037*, 173–186. [CrossRef]

79. Chen, H.Q.; Tannous, M.; Veluthakal, R.; Amin, R.; Kowluru, A. Novel roles for palmitoylation of Ras in IL-1 beta-induced nitric oxide release and caspase 3 activation in insulin-secreting beta cells. *Biochem. Pharmacol.* **2003**, *66*, 1681–1694. [CrossRef]

80. Tannous, M.; Amin, R.; Popoff, M.R.; Fiorentini, C.; Kowluru, A. Positive modulation by Ras of interleukin-1beta-mediated nitric oxide generation in insulin-secreting clonal beta (HIT-T15) cells. *Biochem. Pharmacol.* **2001**, *62*, 1459–1468. [CrossRef]

81. Tannous, M.; Veluthakal, R.; Amin, R.; Kowluru, A. IL-1beta-induced nitric oxide release from insulin-secreting beta-cells: Further evidence for the involvement of GTP-binding proteins. *Diabetes Metab.* **2002**, *28*, 3S78–3S84, discussion 73S108–73S112. [PubMed]

82. Sidarala, V.; Veluthakal, R.; Syeda, K.; Vlaar, C.; Newsholme, P.; Kowluru, A. Phagocyte-like NADPH oxidase (Nox2) promotes activation of p38MAPK in pancreatic β-cells under glucotoxic conditions: Evidence for a requisite role of Ras-related C3 botulinum toxin substrate 1 (Rac1). *Biochem. Pharmacol.* **2015**, *95*, 301–310. [CrossRef] [PubMed]

83. Hernandez-Perez, M.; Chopra, G.; Fine, J.; Conteh, A.M.; Anderson, R.M.; Linnemann, A.K.; Benjamin, C.; Nelson, J.B.; Benninger, K.S.; Nadler, J.L.; et al. Inhibition of 12/15-Lipoxygenase protects against β-Cell oxidative stress and glycemic deterioration in mouse models of type 1 diabetes. *Diabetes* **2017**, *66*, 2875–2887. [CrossRef] [PubMed]

84. Ma, K.; Xiao, A.; Park, S.H.; Glenn, L.; Jackson, L.; Barot, T.; Weaver, J.R.; Taylor-Fishwick, D.A.; Luci, D.K.; Maloney, D.J.; et al. 12-Lipoxygenase inhibitor improves functions of cytokine-treated human islets and type 2 diabetic islets. *J. Clin. Endocrinol. Metab.* **2017**, *102*, 2789–2797. [CrossRef] [PubMed]

85. Jastroch, M. Unraveling the molecular machinery that promotes pancreatic β-cell dysfunction during oxidative stress: Focus on "Phagocyte-like NADPH oxidase promotes cytokine-induced mitochondrial dysfunction in pancreatic β-cells: Evidence for regulation by Rac1". *Am. J. Physiol. Regul. Integr. Comp. Physiol.* **2011**, *300*, R9–R11. [CrossRef]

Publisher's Note: MDPI stays neutral with regard to jurisdictional claims in published maps and institutional affiliations.

 *metabolites*

Review

Islet Health, Hormone Secretion, and Insulin Responsivity with Low-Carbohydrate Feeding in Diabetes

Cassandra A. A. Locatelli [1,2] and Erin E. Mulvihill [1,2,3,4,*]

[1] Energy Substrate Laboratory, The University of Ottawa Heart Institute, 40 Ruskin Street, H-3229A, Ottawa, ON K1Y 4W7, Canada; cloca063@uottawa.ca

[2] Department of Biochemistry, Microbiology and Immunology, The University of Ottawa, Faculty of Medicine, 451 Smyth Rd, Ottawa, ON K1H 8L1, Canada

[3] Montreal Diabetes Research Centre CRCHUM-Pavillion R, 900 Saint-Denis-Room R08.414, Montreal, QC H2X 0A9, Canada

[4] Centre for Infection, Immunity and Inflammation, The University of Ottawa, 451 Smyth Rd, Ottawa, ON K1H 8M5, Canada

* Correspondence: emulvihi@uottawa.ca; Tel.: +1-(613)-696-7221

Received: 1 October 2020; Accepted: 7 November 2020; Published: 11 November 2020

Abstract: Exploring new avenues to control daily fluctuations in glycemia has been a central theme for diabetes research since the Diabetes Control and Complications Trial (DCCT). Carbohydrate restriction has re-emerged as a means to control type 2 diabetes mellitus (T2DM), becoming increasingly popular and supported by national diabetes associations in Canada, Australia, the USA, and Europe. This approval comes from many positive outcomes on HbA1c in human studies; yet mechanisms underlying their success have not been fully elucidated. In this review, we discuss the preclinical and clinical studies investigating the role of carbohydrate restriction and physiological elevations in ketone bodies directly on pancreatic islet health, islet hormone secretion, and insulin sensitivity. Included studies have clearly outlined diet compositions, including a diet with 30% or less of calories from carbohydrates.

Keywords: low carbohydrate diet; ketogenic diet; insulin sensitivity; pancreas; islet of Langerhans

---

1. Introduction

Diabetes mellitus (DM) is a life-altering condition affecting a rapidly growing population worldwide. Type 2 (T2) DM accounts for the majority of this increase and is characterized by hyperglycemia resulting from insufficient endogenous insulin secretion due to beta cell dysfunction or hyperinsulinemia and impaired tissue response to insulin [1]. Appropriately integrated hormonal signals are paramount to initiating proper fasted and post-prandial responses to control glycemia. Both glucose and active incretin peptides stimulate membrane depolarization of pancreatic beta cells and subsequent insulin secretion [2,3]; however, overstimulation of beta cells can lead to impaired secretory function and apoptosis [4]. Thus, modulation of dietary carbohydrate has long been considered for the treatment of DM. Notably, extreme carbohydrate restriction was commonly advised to patients with diabetes before the discovery of insulin [5]. However, recommendations by leading national diabetes organizations have not remained consistent with this initial paradigm, with fat restriction being commonly recommended from the 1970s until recently [6,7]. Since 2017, Diabetes Canada, Diabetes UK, Diabetes Australia, and the American Diabetes Association have approved the use of low-carbohydrate diets (LCD) and ketogenic diets (KD) with support from a physician to manage T2DM [8,9].

Restriction of dietary carbohydrates has become increasingly popular, specifically in T2DM, to improve glycemic control. Increasing the time spent in an optimum glucose range (~4–7 mM) [10] is an important clinical target since the Diabetes Control and Complications Trial that demonstrated that even small deviations in glycemia throughout the day significantly increases risks of secondary diabetic complications [11]. It is understood that beta cells are sensitive to apoptotic signals and oxidative stress in conditions which are exacerbated by overnutrition and obesity [12,13]. Insensitivity to insulin by other metabolic tissues can promote hyperinsulinemia, beta cell hyperplasia, and contribute to beta cell failure. Additionally, hepatic insulin resistance also contributes to hyperglycemia by failing to suppress glucose production [14,15]. Insulin insensitivity, lipid accumulation in the pancreas, and elevated circulating insulin concentrations have been reproducibly demonstrated as outcomes of a Western-style obesogenic diet in animal models [16–19]. T2DM is a heterogeneous condition; however, particularly in the early stages, it is postulated that an LCD may reduce metabolic strain, oxidative stress, and proinflammatory conditions to positively impact beta cell health. However, this has been incompletely demonstrated.

The KD is a very-LCD, typically less than 10% of kcal from carbohydrates, which can induce the fasted-like state of ketosis. During ketosis, excess acetyl-CoA from the catabolism of lipids by hepatic beta-oxidation can be used for ketogenesis, the production of ketone bodies acetoacetate, beta-hydroxybutyrate (BHB), and acetone. Extra-hepatic tissues can take up ketone bodies for ketolysis and generation of ATP through the tricarboxylic acid cycle and electron transport chain [20]. This diet has been historically used to treat children with intractable epilepsy [21]. Further, supplementing exogenous ketones while consuming other diets has also been explored for potential metabolic benefits [22]. These studies provide insight into the effects of ketone bodies independent of the reduced insulin requirement by carbohydrate restriction of KD. It is currently unclear whether the ability of a diet to achieve or maintain ketosis plays a large role in the improved glucose regulation seen with KDs in patients living with DM.

LCD and KDs have shown clinical benefit through reduced need for DM medication in many patients with T2DM and have even been recommended by clinicians as a first approach [23]. However, large-scale, well-controlled trials in patients with T2DM are limited and preclinical research to determine the mechanisms of observed improvements in human trials remains largely controversial. Further, these studies are confounded by the effects of weight loss, appetite suppression, and differing dietary composition of the intervention [24,25].

The aim of this review is to discuss the effects of LCDs, KDs, and dietary supplementation with exogenous ketones on DM as it pertains to: (1) physiology of insulin-secreting beta cells, (2) secretion of islet hormones insulin and glucagon, and (3) sensitivity of other tissues to insulin.

For the purposes of this review, LCDs are defined as having less than 30% of calories from carbohydrates and diets referred to as KD have shown evidence of ketosis through significantly elevated circulating ketone bodies (See Table 1 for all diet definitions). These definitions were used for consistency between rodent, non-human primate, and human studies. Studies included had clearly outlined diet compositions and used LCDs and KDs that are calorically unrestricted, isocaloric to previous eating patterns, or without intentional caloric restriction past control diets. Pubmed search terms included "low carbohydrate islet", "ketogenic diet islet" "low carbohydrate diabetes" "ketogenic diet diabetes". The role of ketones in cancer, including pancreatic cancer have been reviewed elsewhere [26,27].

Table 1. Dietary definitions used for the purpose of this review. Percentages are percent of kilocalories.

Diet			Carbohydrate	Fat	Protein
CRD	Carbohydrate-restricted diet	A diet which intends to decrease carbohydrate consumption	<40%	>30%	4–60%
LCD	Low-carbohydrate diet	A CRD with less than 30% of kcal from carbohydrates without evidence of elevated ketone bodies	<30%	30–95%	4–60%
KD	Ketogenic diet	An LCD with elevated ketone bodies but some dietary carbohydrate and typically low protein	<10%	>70%	4–20%
CFD	Carbohydrate-free diet	A diet containing no carbohydrates (preclinical)	0%	8–88%	12–83%

2. Islet Health and Survival

Progressive beta cell dysfunction and beta cell death are key features of T2DM. Beta cells compensate for insulin insensitivity through hyperplasia and related hyperinsulinemia before a dramatic loss of beta cells. This leads to insufficient insulin production, impaired glucose tolerance, and exogenous insulin dependence [28–30]. Both ectopic lipid accumulation and hyperglycemia induced by Western diet feeding have been demonstrated to promote beta cell toxicity and death in rodents [16,31]. Whether the same pathways are engaged in patients is a matter of debate [32]. Although many KD and LCD studies focus on weight loss and glycemia, histological analysis of pancreatic islet composition in animal models provides important insight into the health and survival of insulin and glucagon secreting beta and alpha cells which remain unknown in human trials. These studies may provide insight into the mechanisms behind the need for altered insulin or anti-hyperglycemic medication which is characteristic of patients with diabetes in trials of LCDs.

2.1. Non-Obese, Non-Diabetic Animal Studies

Interestingly, few positive impacts have been reported on pancreatic islet health in rodents fed LCD and KDs in studies using non-obese, non-diabetic models (Figure 1). C57BL/6 mice fed a KD for 22 weeks showed a greater than 50% decrease in alpha and a ~30% decrease in beta cell mass compared to chow-fed mice without changes to islet density, suggesting increased size of the remaining cells [33]. Similarly, male Wistar rats fed LCD or KD for 4 weeks had decreased pancreas and beta cell volume compared to chow-fed rats, even when normalized for body weight [34]. A recent study of young male and female C57BL/6 mice on three different LCDs for 12 weeks found that only mice on the 1% carbohydrate LCD (0 kcal sucrose) had similar beta cell proliferation (% Ki-67 positive cells) rate as low-fat diet controls. In males, both 20% carbohydrate LCDs had increased proliferation, associated with hyperplasia, whereas only the 20% carbohydrate diet with higher sucrose (775 kcal vs. 275 kcal) was significantly elevated in females, despite identical macronutrients. There were no differences in beta cell mass between diet groups in females, but male mice on the 1% carbohydrate LCD had significantly reduced beta cell mass than even low-fat controls, despite similar proliferation rate [35]. In contrast, studies using obese and/or diabetic models did not report differences between non-obese, non-diabetic control diet groups [36–38].

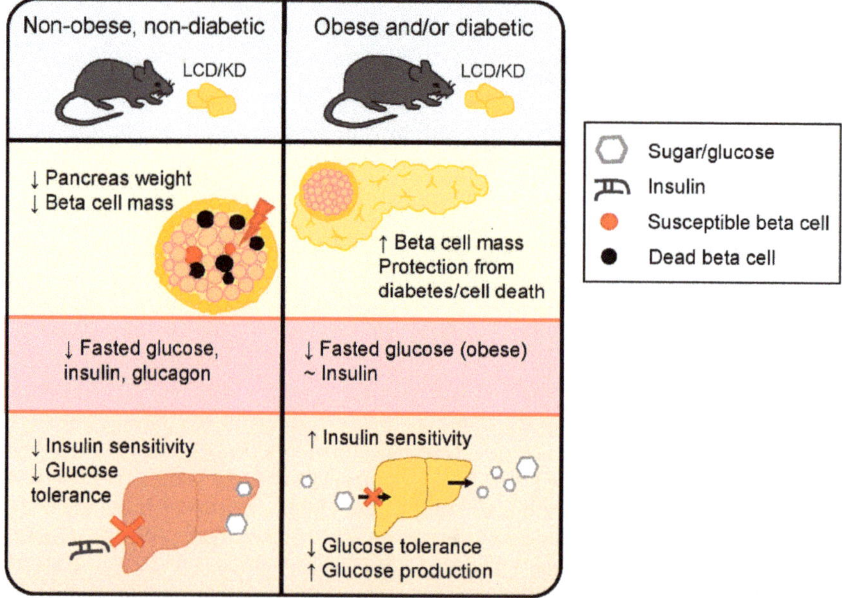

Figure 1. Extreme carbohydrate-restricted diets may confer some benefits in obese and diabetic rodents compared to healthy rodents. Some reports of ketogenic diet (KD) and carbohydrate-free diets (CFDs) demonstrate protection against the onset of diabetes with improved beta cell survival or proliferation after streptozotocin treatment. Low-carbohydrate diets (LCDs) improve fasted and random glycemia but worsen insulin responsivity during hyperinsulinemic euglycemia clamp studies and increase glycemia during glucose tolerance tests.

2.2. Animal Models of Diabetes and/or Obesity

In two studies by the same group, rats with streptozotocin-induced diabetes consuming 10% carbohydrate LCDs had more beta cells than rats fed a high-carbohydrate diet. In the study with a higher fat and lower protein LCD (60% fat, 30% protein), rats had less beta cells than chow-fed controls, suggesting that this diet was inferior for islet mass preservation compared to typical healthy feeding. [36,37]. More strikingly, streptozotocin-treated rats fed the very high protein LCD (60% protein 30% fat) did not show signs of streptozotocin-induced beta cell death and islet morphology perturbations, unlike in both chow and high-carbohydrate-fed rats [36]. Notably, diabetes was induced by streptozotocin at different times in these two studies. The higher fat LCD experiment induced diabetes at the onset of dietary intervention whereas in the experiment with higher protein LCD, dietary intervention began 8 weeks prior to streptozotocin treatment. In both studies, chow and high carbohydrate groups increased their food intake after streptozotocin treatment, but LCD-fed rats did not [36,37].

In the diabetic susceptible mouse model *db/db*, Mirhashemi and colleagues found that a carbohydrate-free diet (CFD) preserved beta cells and expression of glucose transporter (GLUT2) after 22 weeks of dietary intervention. Both the standard chow and obesogenic HFD feeding resulted in diminished beta cell count [39]. Similarly, a 1983 study on the role of carbohydrate in *db/db* mice found that the CFD-fed mice showed some islet hyperplasia and hypertrophy, but significantly reduced islet atrophy and improved survival compared to those on diets of varied carbohydrates (8–60%) [40]. These results are corroborated in New Zealand Obese (NZO) mice fed a CFD for 22 weeks [41], but not NZO mice fed a very LCD (6% kcal from sucrose) for 9 weeks; here, Lamont et al. found no changes to pancreas weight, islet density, islet size, or beta cell mass [42]. These results suggest a benefit on islet health in rodents with a total lack of carbohydrate, not only carbohydrate restriction, as higher sucrose (compared to 2% in standard diet) is detrimental to beta cell health [40]. However, this raises questions on the durability of these effects upon the reintroduction of carbohydrates after a CFD. After 18 weeks on a CFD, Kluth et al. fed NZO mice a diet with 32% of kcal from carbohydrates or continued the CFD. When refed carbohydrates, mice quickly developed diabetes demonstrated by loss of insulin-positive cells and increased caspase 3 expression [43]. It is important to note that the carbohydrate refeeding diet was proportionately closer to a typical HFD (51.4% of kcal from fat, 32.4% kcal from carbohydrate), which is known to be metabolically deleterious. However, a study on chow refeeding in rats after 8 weeks of KD demonstrated hyperphagia and enhanced weight gain compared to mice consistently fed a chow diet. Islet and pancreas analyses were not performed [44]. Finally, in the *ob/ob* genetic model of obesity, mice fed an LCD had reduced beta cell mass compared to *ob/ob* chow-fed mice, similar to levels in wildtype chow fed mice, indicating hyperplasia in the chow-fed *ob/ob* mice which is attenuated by the LCD [38].

3. Islet Hormone Secretion

Constant glucose homeostasis as a result of appropriate secretion of metabolism-regulating islet hormones or treatment with diabetes medication is important for daily metabolism, avoidance of hyper- and hypoglycemic events, and is crucial for patients with diabetes to avoid long term complications [11]. Physiological regulation of glucose by the endocrine pancreas occurs through the integration of nutrient signals with a series of electrical gradients and molecular processes which result in the secretion of glucoregulatory hormones, insulin and glucagon, from the islets of Langerhans. Investigating the effects of LCD, KD, and their metabolites on glycemic load and islet secretory capacity compared to standard diets is paramount to understanding their role in DM and its comorbidities.

3.1. Cell Studies

The canonical mechanism of insulin secretion occurs in response to glucose; however, in vitro studies in isolated rat and human islets have shown that exposure to BHB in addition to glucose further increases insulin secretion but does not have a significant effect alone [45–49]. Further, while acetoacetate, BHB, and monomethyl succinate had no significant effects on insulin secretion in INS-1 cells alone, incubation of either acetoacetate or BHB with monomethyl succinate induced significant insulin secretion [48]. Similarly, monomethyl succinate and BHB together also stimulated insulin secretion in isolated rat islets, though neither alone [48]. Interestingly, one 1995 study shows that treatment of human islets for 48 h with BHB at concentrations seen in patients with uncontrolled type 1 DM impairs insulin secretion in the presence of high glucose media [50]. Conversely, a recent study found that supplementing high-glucose media with medium chain triglycerides improved glucose-stimulated insulin secretion in isolated beta cells from aged rats [51]. Chromic (72 h) treatment with medium chain triglycerides or BHB also increased glucose-stimulated insulin secretion in the INS1E beta cell line [51] (Figure 2). Proinsulin biosynthesis in islets from *ob/ob* mice, in significant contrast to glucose, was unchanged in the presence of ketone bodies [52]. A number of studies have evaluated the expression and activity of ketone body transport and utilization enzymes within islet cell lines, primary pancreatic tissue [53,54] and within mouse islets [55]. Additionally, oxidation rates in islets isolated from *ob/ob* mice (6–8 months old) determined conversion of ketone bodies to CO_2 occurs at significant rates albeit much lower than other tissues, including the kidney [56]. However, surprisingly few studies have explored the physiological adaptation of islets to carbohydrate restriction, nutritional ketosis, or consumption of ketone esters on the dynamics of hormonal secretion and metabolism within islet cell populations.

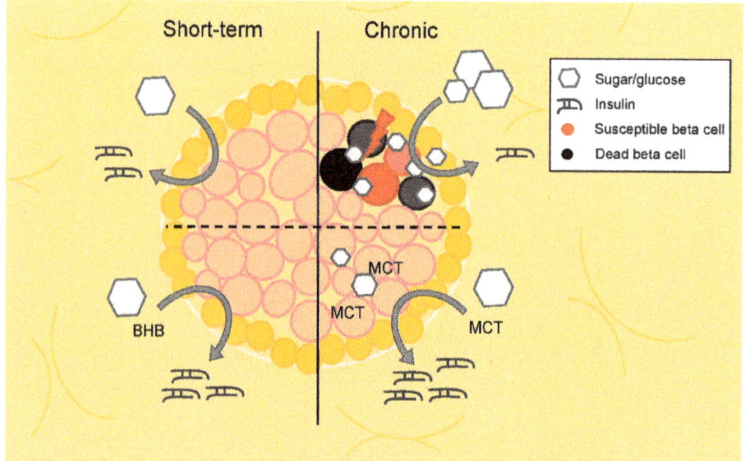

Figure 2. Short-term and chronic effects of metabolites on pancreatic beta cell insulin secretion. In vitro investigations in isolated islets reveal an additive effect of BHB and glucose on insulin secretion, but no insulin secretion with BHB alone. While long-term treatment with high glucose can lead to reduced glucose-stimulated insulin secretion, some evidence suggests increased insulin secretion in response to high glucose following chronic (72 h) treatment with ketones and medium chain triglycerides. BHB: beta-hydroxy butyrate; MCT: medium chain triglycerides.

3.2. Healthy Animal Models

In non-obese, non-diabetic rodent studies, LCD and KD feeding often results in significantly lower fasted glucose than rodents on a chow diet [34,57–60], or no different than chow or low-fat diet [33,35,61,62] and improved compared to obesogenic HFDs. Fed state glucose was also decreased in KD compared to chow-fed mice [61]; however, it was not different than chow-fed rats [37,63]. When challenged with systemic (insulin tolerance test) or neural hypoglycemia, mice on the KD for 7 days had significantly lower glucagon and blood glucose compared to chow-fed mice, indicating reduced protection for hypoglycemic events [60]. Jornayvaz et al. similarly found that mice fed a KD had decreased fasted glucagon compared to chow-fed mice [59], whereas in another study, KD mice had similar fed glucagon to weight matched, chow-fed controls [61]. Fasted insulin has been reported to decrease with KD feeding compared to chow feeding [34,57,59,61,62,64]; however, other studies report that fasted insulin is not different between KD and chow-fed mice, although decreased compared to HFD-fed male mice [35,58]. Interestingly, in the longest running study we evaluated, mice fed a KD for 22 weeks had increased fasted insulin compared to chow-fed mice [33]. Finally, non-human primates fed standard high carbohydrate diet had less insulin and glucose perturbations with a LCD meal compared to high-carbohydrate, low-fat meal [65].

3.3. Animal Models of Obesity and Diabetes

Unlike in healthy rodents, KD feeding in rodent models of obesity and diabetes have more varied fasted glucose outcomes. In *ob/ob* and diet-induced obese mice, fasted glucose was lower in KD-fed mice than in chow-fed mice [38,57,61]. However, fasting glucose was not significantly different from chow-fed rodents in pancreatectomized, obese, or streptozotocin treated rodents fed LCD and KD [63,66,67] and was increased in LCD-fed NZO mice [42]. Unsurprisingly, fed glucose levels in obese and diabetic rodents on LCDs are consistently lower than chow-fed rodents [37,38,42,61,68]. Interestingly, Park et al. report pancreatectomized KD mice had more than doubled average fasted glucagon levels than pancreatectomized chow and exogenous ketone-supplemented groups [66], whereas Badman et al. found decreased fed glucagon levels in KD compared to chow-fed *ob/ob* mice [61].

Unlike in the healthy rodents on LCDs who reproducibly have lower fasted insulin, only one study using a mouse model of obesity (*ob/ob*), had decreased fasted insulin as compared to chow fed mice [61]. Others, using pancreatectomized rats, streptozotocin treated rats, or obese (NZO) mice, showed no change in fasted plasma insulin levels with LCD, KD, or exogenous ketones compared to chow [42,66,67]. Using a CFD, Mirhashemi et al. found that non-fasted insulin was consistent over time in *db/db* obese mice despite a decrease in insulin in HFD-fed mice at 6 weeks and chow-fed mice by the endpoint (22 weeks), suggesting decompensation [39]. Finally, non-fasted insulin levels of obese LCD and KD rodents tended to be decreased compared to higher carbohydrate chow and HFDs [38,57,61,63,68]. Although, one study using NZO mice found increased fed insulin levels after 9 weeks of LCD feeding [42]. In all, LCD and KDs protect rodents from hyperglycemia and large glycemic fluctuations.

3.4. Human Studies

In most cases, human studies on LCD and KD feeding report significantly reduced fasted and 24 h insulin and increased fasted, non-fasted, and 24 h glucagon compared to control groups or baseline measurement [69–74]. Interestingly, Samaha and colleagues found decreased insulin levels in severely obese participants currently without DM medications, but not with those using DM medications [75]. In contrast, a 52-week study comparing LCD and high carbohydrate hypocaloric diet found no difference in serum insulin between groups [76]. Similarly, a 2-week crossover study found no changes to insulin or glucagon at basal insulin before hyperinsulinemic clamp [77].

Measurement of HbA1c provides a much more complete picture of glycemia than one-time fasted or fed glucose measurement and is indicative of inadequate insulin secretion or glucose uptake over time. Impressively, studies in patients with T2DM on LCD and KD both randomized and non-randomized, controlled and uncontrolled trials, from 2 weeks to 44 months reproducibly resulted in decreased HbA1c [69,71,72,75,78–86]. Similarly, case studies of varying lengths tended to also find improvements in HbA1c after 20 weeks [87]. Controlled studies have found this improvement in HbA1c on LCD and KDs compared against diets with the majority of energy coming from carbohydrates, diets with low glycemic index and calorie restriction, and caloric restriction alone [69,79,81,84]. In addition, continuous care regimens using KDs have be shown to greatly improve HbA1c in patients with T2DM compared to current standard T2DM care [88,89]. Strikingly, many participants in the continuous care group with KD had reversal (53.5%) or remission (17.6%) of their T2DM [88]. While continuous care interventions introduce the confounding variables of extra support and resources, these studies offer insight into the potential for improved care in conjunction with the benefits demonstrated by other trials of CRDs alone. Conversely, one 6-month study comparing a LCD and a low-fat Diabetes UK recommended diet, HbA1c only trended towards improvement in the LCD compared to the control diet among participants with DM ($p = 0.06$) [75]. This study saw significantly decreased weight compared to the low-fat group, and the decrease in fasting glucose that was seen in diabetic patients was no longer significant after controlling for weight loss. In a 24 weeks study, Dyson and colleagues found HbA1c was only significantly decreased with LCD feeding compared to baseline and not the control low-fat diet when both patients with and without DM are considered and not DM alone, despite increased weight loss with LCD compared to the low-fat diet [25]. Similarly, a 52-week study found that improvements in HbA1c were not different than those of the high carbohydrate group [76]. Remarkably, Yancy et al. found that the decrease in HbA1c was greater than that which is predicted by the weight loss and Vernon et al. found that HbA1c improvement was observed in patients with and without weight loss, alike [78,80]. Additionally, by nature of carbohydrate restriction and the positive outcomes above, human studies on the CRDs almost always result in decreased reliance on anti-hyperglycemic agents and/or lowering of insulin dosage [69,71,75–77,79–81,83,84,90], and on occasion even significantly more than calorie-restricted and low glycemic index diets [69,79].

4. Insulin Sensitivity

Insulin insensitivity is an important hallmark of T2DM as a lack of tissue responsiveness to insulin remains a barrier to stabilizing glycemia. In rodent studies, there are a number of accepted methods for testing whole body and tissue-specific insulin sensitivity. The gold standard for insulin sensitivity measurement is the hyperinsulinemic-euglycemic clamp; however, glucose and insulin tolerance tests are very widely used. Additionally, when dynamic physiological tests are not performed, simple blood draws can be used to measure homeostatic model assessment for insulin resistance (HOMA-IR) and quantitative insulin sensitivity check index (QUICKI). While studies demonstrate consistencies between methods in humans, such comparisons between methods in rodents are not as appropriate [91–93].

4.1. Non-Obese, Non-Diabetic Mouse Models

In both C57BL6 mice fed a KD for 3 days and 5 weeks and Wistar rats fed LCD and KD for 3 weeks, hyperinsulinemic-euglycemic clamp studies show reduced hepatic insulin sensitivity and inability to suppress endogenous glucose production [34,58,59]. Additionally, in two of those studies, rodents had impaired uptake of glucose during the clamp experiment [34], specifically in heart and brown adipose tissue [59]. Most insulin tolerance tests (ITT) on healthy mice fed CRDs report lower or the same glycemia compared to a standard chow diet [33,60,62,94]. Notably, two KD-fed mice in one study became hypoglycemic during the ITT and required rescue with dextrose solution [62]. However, the shortest study included, with only 3 days on KD, reported increased area under the curve of glucose compared to both a chow and obesogenic HFD during the ITT [58]. With only three days on the diet it is possible that this represents a metabolic transition phase rather than an established state [58]. In non-obese, non-diabetic rats, three studies found that both LCD and KD impaired insulin tolerance through elevated glucose during the ITT compared to chow-fed rats [34,63,64]. Similarly, Kinzig and colleagues found that feeding a liquid high carbohydrate meal to rats on the LCD significantly increased circulating insulin and glucose for 2 h after feeding compared to chow-fed rats [64]. Non-validated measurements of insulin sensitivity in rodents, QUICKI and HOMA-IR, were improved in studies of LCD and KD in healthy wildtype mice [59,62]. Studies in healthy rodents agree that CRDs impair glucose tolerance as assessed by oral and intraperitoneal glucose tolerance test [33–35,38,58,62,64,94,95]. Interestingly, a recent study revealed that in healthy mice on diets with fixed 25% protein, those with fat content of 41.7% and above were all glucose intolerant; this phenomenon was not seen with 10% fixed protein, despite that being much closer to a typical KD [96].

4.2. Mouse Models of Obesity and Diabetes

Similar to healthy rodents, rodent models of diabetes and obesity fed CRDs had poorer outcomes during hyperinsulinemic-euglycemic clamp experiments than chow-fed controls. NZO mice fed a CFD fared similarly to HFD-fed mice with decreased glucose infusion rate to achieve euglycemia, decreased glucose utilization, and decreased suppression of hepatic glucose production compared to chow-fed mice [41]. Pancreatectomized rats on KD also had lower glucose infusion to achieve euglycemia and increased hepatic glucose output. No differences were found in glucose uptake between KD, chow, and BHB injection groups [66]. Interestingly, unlike KD rats, rats administered exogenous BHB had decreased hepatic glucose output compared those on chow [66]. During hyperglycemic clamp, BHB injected and KD rats had reduced first phase insulin secretion while only BHB injected rats had reduced second phase insulin [66]. Both KD and BHB groups had reduced glucose infusion rate and insulin sensitivity, measured as milligrams of glucose per kg of body weight over time [66]. This speaks to a potential benefit of increased fat content of KD over exogenous BHB which may allow for adequate second phase insulin secretion. During ITTs, obese (*ob/ob*) mice had lower blood glucose than chow-fed controls, indicating better insulin sensitivity; however, these mice again were glucose intolerant. Mice with diet-induced obesity switched to an LCD were also glucose intolerant but they had similar insulin tolerance to those switched to chow [35]. Similarly, LCD-fed rats treated with streptozotocin had similar ITT response as non-streptozotocin chow fed rats, but their glucose tolerance, while improved at some time points, did not have significantly improved area under the curve (AUC) glucose compared to HFD streptozotocin controls [67]. Genetically obese (*fa/fa*) rats fed an LCD with high protein (34.7% kcal) also had similar insulin tolerance to chow-fed controls, with increased AUC of glucose and insulin during a glucose tolerance test [63]. Conversely, two studies found that KD-fed mice had similar glucose tolerance to chow-fed mice and improved glucose tolerance compared to HFD-fed mice in models of diet and genetically (*ob/ob*) induced obesity [57,61]. Insulin sensitivity, as determined by QUICKI, was improved in obese (*ob/ob*) mice [61].

4.3. Human Studies

Two human studies have assessed insulin sensitivity using hyperinsulinemic-euglycemic clamps in LCD and KD fed participants with T2DM. In contrast to rodent studies, neither short term, small population study had worsened insulin sensitivity with LCD and KD [71,77]. Notably, KD fed participants in one study required significantly more glucose to maintain euglycemia than the same participants did on their prior diets [71] (Figure 3). This inpatient study lacked a control group; therefore, these improvements may be a reflection on the poor diet they consumed prior to the study but may also serve as a more realistic comparison for the average person with T2DM eating a Western-style diet [71]. Allick et al. compared their findings against a eucaloric, high carbohydrate diet and found no differences between groups during the clamp study despite significantly worsened hepatic insulin sensitivity in a similar study done on non-diabetic participants [77,97]. Additionally, in a study of non-diabetic, obese adults over 60, participants on the LCD for 8 weeks had improved insulin sensitivity during hyperinsulinemic-euglycemic clamp compared to baseline, while the low-fat group saw no significant change. The difference between diet groups was not significant [73]. In mixed meal tests, Rosenbaum et al. found that 17 healthy participants fed KD versus a baseline 50% carbohydrate diet had worse insulin sensitivity with elevated AUC glucose during ketogenic and carbohydrate-based meals and elevated AUC insulin during the carbohydrate-based meals [74]. Further investigation is required to assess insulin sensitivity on LCDs in the context of T2DM.

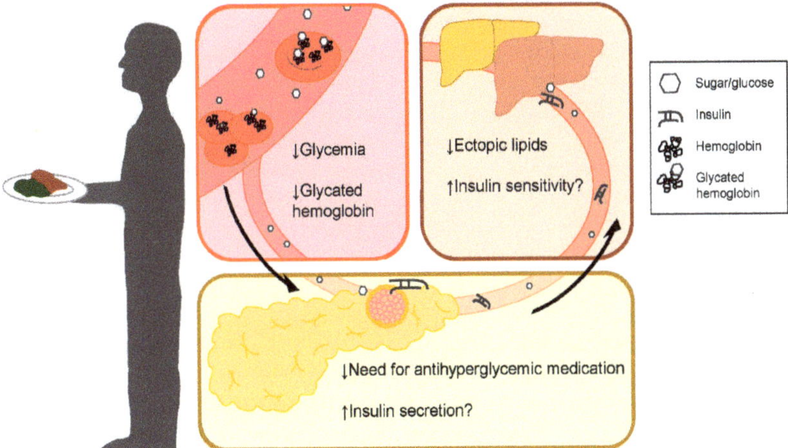

Figure 3. Low-carbohydrate diets reduce the need for anti-hyperglycemic agents in type 2 diabetes. Participants on LCDs with T2D have decreased glycemia, glycated hemoglobin, and improved blood glucose regulation; thus, require less exogenous insulin and other anti-hyperglycemic medications and may improve insulin sensitivity through restored insulin secretion in some patients.

5. Paracrine Regulation and Metabolic Crosstalk

Paracrine regulation of islet hormone secretion may also be of importance to the LCD and diabetes literature. The incretins glucagon-like peptide 1 (GLP-1) and glucose-dependent insulinotropic polypeptide (GIP) are secreted by L and K cells in the intestine in response to nutrients [2]. Secreted, active GLP-1 and GIP can stimulate beta cells of the pancreas to potentiate glucose-stimulated insulin secretion, with GLP-1 having the added effect of inhibiting glucagon secretion from alpha cells [2]. However, the incretins are quickly cleaved and inactivated by dipeptidyl peptidase-4 (DPP4) and cleared through the kidney [98]. Both DPP4 inhibitors and GLP-1 receptor agonists are currently in use as DM medications [99,100]. One study reported that normal weight participants taking a ketone ester drink had significantly lower total GLP-1 two to four hours after drink administration compared to those taking a dextrose drink [101]. Similarly, Wallenius and colleagues reported that ketone bodies reduce GLP-1 secretion in primary jejunal culture [102]. Alternatively, a small, uncontrolled study reported increased fasted GIP in healthy patients on a KD but no changes to GLP-1 [74]. The differences here may be due to the vastly different diets; very low-calorie diets and ketone ester drinks do not provide the extreme fat content of a calorically unrestricted KD. Dietary fatty acids can signal through G protein coupled receptors in the intestine which can potentiate GLP-1 and GIP secretion [103–105]. Further studies are required to understand the relationship between incretin secretion and KD for any therapeutic benefits.

6. Discussion and Conclusions

In summary, the preclinical literature supports the concept that islet health may be positively impacted by CRD in states of metabolic disease rather than in the healthy state. Much of the literature on dietary modulation on pancreas and islet health centers on beta cell destruction; as such, the diet regime is extreme with many studies using diets free of carbohydrates, which clearly limits its translational value. Understanding the functional impact of different dietary regimes on the energy sources presented to the beta cell becomes increasingly relevant as we learn more on the importance of lipid storage within the pancreas, integrity of mitochondria and oxidative metabolism on insulin secretion and the development of diabetes [106,107]. Additionally, it is clear, given the significant effects of diets with a high percentage of fat reported on body weight, food intake, and energy expenditure [108], that metabolic cross-talk from insulin sensitive tissues including adipose tissue, liver, and skeletal muscle may also very dramatically influence the islet and metabolic outcomes associated with the feeding of LCD or KD. CRDs with differing composition should be studied more closely and rigorously to address the mechanisms underlying reported inconsistencies and the translational value of these studies. Additionally, exercise can play an important role in the management of diabetes and other metabolic dysfunction [109–111]. There is some evidence to suggest that an LCD or KD may impair exercise ability [112–115], although this is not consensus and may depend on exercise type [116–118]. Further studies on the role of ketosis and carbohydrate restriction on exercise capacity in patients with diabetes is warranted.

Carbohydrate restriction (<30% kcal) has demonstrated benefits in human studies of T2DM with regulating glycemia and reducing the need for DM medications. Interventional studies in patients with T2DM examining changes in glycemia, insulin sensitivity, and beta cell function provide valuable insight into the mechanisms underlying the consistency by which LCD or KD may improve HbA1c levels. The value of improved glycemic control is evident given the significant role of the length of exposure to hyperglycemia in the development of complications. It is currently unclear how factors which influence the heterogeneity of T2DM including age, race, and genetics [119] influence the ability of carbohydrate restriction to resolve morbidity. In addition, there is substantial debate about the heterogeneity in the sources of fat present in the diet [120–122] which may also influence the consistency of these outcomes. Further, like any dietary restriction or modulation, adherence is a major obstacle to overcome when informing clinical practice and evaluating long-term outcomes. While carbohydrate consumption was decreased in a study aiming to explore an LCD, participants randomized to the LCD group did not have significantly increased fat intake compared to controls, rather decreased caloric intake [123]. A 2018 review also highlights the difficulty of adhering to a KD [124]; thus, studying the transition from a more restrictive KD to an LCD for the long term may be more clinically relevant.

Author Contributions: Writing—original draft preparation, C.A.A.L. and E.E.M.; writing—review and editing, C.A.A.L. and E.E.M.; funding acquisition, E.E.M. All authors have read and agreed to the published version of the manuscript.

Funding: This research was funded by Canadian Institutes of Health Research Grants, ARJ-162628 and Project Grant, 156136 to E.E.M. E.E.M. is the recipient of a Diabetes Canada New Investigator Award.

Acknowledgments: The authors would like to acknowledge Nadya Morrow and Ilka Lorenzen-Schmidt for insightful contributions.

Conflicts of Interest: The Mulvihill lab receives funding from the Merck IISP program for pre-clinical studies unrelated to this work. The other authors have no disclosures.

References

1. Esser, N.; Utzschneider, K.M.; Kahn, S.E. Early Beta Cell Dysfunction vs Insulin Hypersecretion as the Primary Event in the Pathogenesis of Dysglycaemia. *Diabetologia* **2020**, *63*, 2007–2021. [CrossRef] [PubMed]

2. Baggio, L.L.; Drucker, D.J. Biology of Incretins: GLP-1 and GIP. *Gastroenterology* **2007**, *132*, 2131–2157. [CrossRef] [PubMed]

3. Rorsman, P.; Eliasson, L.; Renström, E.; Gromada, J.; Barg, S.; Göpel, S. The Cell Physiology of Biphasic Insulin Secretion. *News Physiol. Sci.* **2000**, *15*, 72–77. [CrossRef] [PubMed]

4. Grill, V.; Björklund, A. Overstimulation and β-Cell Function. *Diabetes* **2001**, *50*, 48–50. [CrossRef]

5. Westman, E.C.; Yancy, W.S.; Humphreys, M. Dietary Treatment of Diabetes Mellitus in the Pre-Insulin Era (1914–1922). *Perspect. Biol. Med.* **2006**, *49*, 77–83. [CrossRef]

6. Moran, M. The Evolution of the Nutritional Management of Diabetes. *Proc. Nutr. Soc.* **2004**, *63*, 615–620. [CrossRef]

7. Nuttall, F.Q.; Brunzell, J.D. Principles of Nutrition and Dietary Recommendations for Individuals with Diabetes Mellitus: 1979. *Diabetes Care* **1979**, *28*, 520–523.

8. Article, S. Diabetes Canada Position Statement on Low-Carbohydrate Diets for Adults with Diabetes: A Rapid Review. *Can. J. Diabetes* **2020**, *44*, 295–299.

9. Evert, A.B.; Dennison, M.; Gardner, C.D.; Timothy Garvey, W.; Karen Lau, K.H.; MacLeod, J.; Mitri, J.; Pereira, R.F.; Rawlings, K.; Robinson, S.; et al. Nutrition Therapy for Adults with Diabetes or Prediabetes: A Consensus Report. *Diabetes Care* **2019**, *42*, 731–754. [CrossRef]

10. Managing Your Blood Sugar. Available online: https://www.diabetes.ca/managing-my-diabetes/tools---resources/managing-your-blood-sugar (accessed on 27 September 2020).

11. Nathan, D.M. The Diabetes Control and Complications Trial/Epidemiology of Diabetes Interventions and Complications Study at 30 Years: Overview. *Diabetes Care* **2014**, *37*, 9–16. [CrossRef]

12. Johnson, J.D.; Luciani, D.S. Mechanisms of Pancreatic β-Cell Apoptosis in Diabetes and Its Therapies. *Adv. Exp. Med. Biol.* **2010**, *654*, 447–462.

13. Butler, A.E.; Janson, J.; Bonner-Weir, S.; Ritzel, R.; Rizza, R.A.; Butler, P.C. β-Cell Deficit and Increased β-Cell Apoptosis in Humans with Type 2 Diabetes. *Diabetes* **2003**, *52*, 102–110. [CrossRef] [PubMed]

14. Rizza, R.A. Pathogenesis of Fasting and Postprandial Hyperglycemia in Type 2 Diabetes: Implications for Therapy. *Diabetes* **2010**, *59*, 2697–2707. [CrossRef] [PubMed]

15. Firth, R.G.; Bell, P.M.; Marsh, H.M.; Hansen, I.; Rizza, R.A. Postprandial Hyperglycemia in Patients with Noninsulin-Dependent Diabetes Mellitus. Role of Hepatic and Extrahepatic Tissues. *J. Clin. Investig.* **1986**, *77*, 1525–1532. [CrossRef] [PubMed]

16. Heydemann, A. An Overview of Murine High Fat Diet as a Model for Type 2 Diabetes Mellitus. *J. Diabetes Res.* **2016**, *2016*. [CrossRef] [PubMed]

17. Paschen, M.; Moede, T.; Valladolid-Acebes, I.; Leibiger, B.; Moruzzi, N.; Jacob, S.; García-Prieto, C.F.; Brismar, K.; Leibiger, I.B.; Berggren, P.O. Diet-Induced b-Cell Insulin Resistance Results in Reversible Loss of Functional b-Cell Mass. *FASEB J.* **2019**, *33*, 204–218. [CrossRef]

18. Matsuda, A.; Makino, N.; Tozawa, T.; Shirahata, N.; Honda, T.; Ikeda, Y.; Sato, H.; Ito, M.; Kakizaki, Y.; Akamatsu, M.; et al. Pancreatic Fat Accumulation, Fibrosis, and Acinar Cell Injury in the Zucker Diabetic Fatty Rat Fed a Chronic High-Fat Diet. *Pancreas* **2014**, *43*, 735–743. [CrossRef]

19. Pinnick, K.E.; Collins, S.C.; Londos, C.; Gauguier, D.; Clark, A.; Fielding, B.A. Pancreatic Ectopic Fat Is Characterized by Adipocyte Infiltration and Altered Lipid Composition. *Obesity* **2008**, *16*, 522–530. [CrossRef]

20. Puchalska, P.; Crawford, P.A. Multi-Dimensional Roles of Ketone Bodies in Fuel Metabolism, Signaling, and Therapeutics. *Cell Metab.* **2017**, *25*, 262–284. [CrossRef]

21. Cooder, H.R. Epilepsy in Children: With Particular Reference to the Ketogenic Diet. *Cal. West. Med.* **1933**, *39*, 169–173.

22. Stubbs, B.J.; Cox, P.J.; Evans, R.D.; Santer, P.; Miller, J.J.; Faull, O.K.; Magor-Elliott, S.; Hiyama, S.; Stirling, M.; Clarke, K. On the Metabolism of Exogenous Ketones in Humans. *Front. Physiol.* **2017**, *8*, 1–13. [CrossRef] [PubMed]

23. Feinman, R.D.; Pogozelski, W.K.; Astrup, A.; Bernstein, R.K.; Fine, E.J.; Westman, E.C.; Accurso, A.; Frassetto, L.; Gower, B.A.; McFarlane, S.I.; et al. Dietary Carbohydrate Restriction as the First Approach in Diabetes Management: Critical Review and Evidence Base. *Nutrition* **2015**, *31*, 1–13. [CrossRef] [PubMed]

24. Johnstone, A.M.; Horgan, G.W.; Murison, S.D.; Bremner, D.M.; Lobley, G.E. Effects of a High-Protein Ketogenic Diet on Hunger, Appetite, and Weight Loss in Obese Men Feeding Ad Libitum. *Am. J. Clin. Nutr.* **2008**, *87*, 44–55. [CrossRef] [PubMed]

25. Dyson, P.A.; Beatty, S.; Matthews, D.R. A Low-Carbohydrate Diet Is More Effective in Reducing Body Weight than Healthy Eating in Both Diabetic and Non-Diabetic Subjects. *Diabet. Med.* **2007**, *24*, 1430–1435. [CrossRef]

26. Weber, D.D.; Aminzadeh-Gohari, S.; Tulipan, J.; Catalano, L.; Feichtinger, R.G.; Kofler, B. Ketogenic Diet in the Treatment of Cancer—Where Do We Stand? *Mol. Metab.* **2020**, *33*, 102–121. [CrossRef]

27. Tan-Shalaby, J. Ketogenic Diets and Cancer: Emerging Evidence. *Fed. Pract.* **2017**, *34*, 37S–42S.

28. Murai, N.; Saito, N.; Kodama, E.; Iida, T.; Mikura, K.; Imai, H.; Kaji, M.; Hashizume, M.; Kigawa, Y.; Koizumi, G.; et al. Insulin and Proinsulin Dynamics Progressively Deteriorate From Within the Normal Range Toward Impaired Glucose Tolerance. *J. Endocr. Soc.* **2020**, *4*. [CrossRef]

29. Quan, W.; Jo, E.K.; Lee, M.S. Role of Pancreatic β-Cell Death and Inflammation in Diabetes. *Diabetes Obes. Metab.* **2013**, *15*, 141–151. [CrossRef]

30. Van Vliet, S.; Koh, H.-C.; Patterson, B.; Yoshino, M.; LaForest, R.; Gropler, R.; Klein, S.; Mittendorfer, B. Obesity Is Associated with Increased Basal and Postprandial β-Cell Insulin Secretion Even in the Absence of Insulin Resistance. *Diabetes* **2020**, *69*, 2112–2119. [CrossRef]

31. Rojas, J.; Bermudez, V.; Palmar, J.; Martínez, M.S.; Olivar, L.C.; Nava, M.; Tomey, D.; Rojas, M.; Salazar, J.; Garicano, C.; et al. Pancreatic Beta Cell Death: Novel Potential Mechanisms in Diabetes Therapy. *J. Diabetes Res.* **2018**. [CrossRef]

32. Weir, G.C. Glucolipotoxicity, β-Cells, and Diabetes: The Emperor Has No Clothes. *Diabetes* **2020**, *69*, 273–278. [CrossRef] [PubMed]

33. Ellenbroek, J.H.; Van Dijck, L.; Töns, H.A.; Rabelink, T.J.; Carlotti, F.; Ballieux, B.E.P.B.; De Koning, E.J.P. Long-Term Ketogenic Diet Causes Glucose Intolerance and Reduced β- and α-Cell Mass but No Weight Loss in Mice. *Am. J. Physiol. Endocrinol. Metab.* **2014**, *306*, 552–558. [CrossRef] [PubMed]

34. Bielohuby, M.; Sisley, S.; Sandoval, D.; Herbach, N.; Zengin, A.; Fischereder, M.; Menhofer, D.; Stoehr, B.J.M.; Stemmer, K.; Wanke, R.; et al. Impaired Glucose Tolerance in Rats Fed Low-Carbohydrate, High-Fat Diets. *Am. J. Physiol. Endocrinol. Metab.* **2013**, *305*, E1059–E1070. [CrossRef] [PubMed]

35. Her, T.K.; Lagakos, W.S.; Brown, M.R.; LeBrasseur, N.K.; Rakshit, K.; Matveyenko, A.V. Dietary Carbohydrates Modulate Metabolic and β-Cell Adaptation to High-Fat Diet-Induced Obesity. *Am. J. Physiol. Endocrinol. Metab.* **2020**, *318*, E856–E865. [CrossRef] [PubMed]

36. Al-Khalifa, A.; Mathew, T.C.; Al-Zaid, N.S.; Mathew, E.; Dashti, H. Low Carbohydrate Ketogenic Diet Prevents the Induction of Diabetes Using Streptozotocin in Rats. *Exp. Toxicol. Pathol.* **2011**, *63*, 663–669. [CrossRef]

37. Al-Khalifa, A.; Mathew, T.C.; Al-Zaid, N.S.; Mathew, E.; Dashti, H.M. Therapeutic Role of Low-Carbohydrate Ketogenic Diet in Diabetes. *Nutrition* **2009**, *25*, 1177–1185. [CrossRef]

38. Tattikota, S.G.; Rathjen, T.; McAnulty, S.J.; Wessels, H.H.; Akerman, I.; Van De Bunt, M.; Hausser, J.; Esguerra, J.L.S.; Musahl, A.; Pandey, A.K.; et al. Argonaute2 Mediates Compensatory Expansion of the Pancreatic β Cell. *Cell Metab.* **2014**, *19*, 122–134. [CrossRef]

39. Mirhashemi, F.; Kluth, O.; Scherneck, S.; Vogel, H.; Kluge, R.; Schürmann, A.; Joost, H.G.; Neschen, S. High-Fat, Carbohydrate-Free Diet Markedly Aggravates Obesity but Prevents β-Cell Loss and Diabetes in the Obese, Diabetes-Susceptible Db/Db Strain. *Obes. Facts* **2008**, *1*, 292–297. [CrossRef]

40. Leiter, E.H.; Coleman, D.L.; Ingram, D.K.; Reynolds, M.A. Influence of Dietary Carbohydrate on the Induction of Diabetes in C57BL/KsJ-Db/Db Diabetes Mice. *J. Nutr.* **1983**, *113*, 184–195. [CrossRef]

41. Jürgens, H.S.; Neschen, S.; Ortmann, S.; Scherneck, S.; Schmolz, K.; Schüler, G.; Schmidt, S.; Blüher, M.; Klaus, S.; Perez-Tilve, D.; et al. Development of Diabetes in Obese, Insulin-Resistant Mice: Essential Role of Dietary Carbohydrate in Beta Cell Destruction. *Diabetologia* **2007**, *50*, 1481–1489. [CrossRef]

42. Lamont, B.J.; Waters, M.F.; Andrikopoulos, S. A Low-Carbohydrate High-Fat Diet Increases Weight Gain and Does Not Improve Glucose Tolerance, Insulin Secretion or β-Cell Mass in NZO Mice. *Nutr. Diabetes* **2016**, *6*, e194. [CrossRef] [PubMed]

43. Kluth, O.; Mirhashemi, F.; Scherneck, S.; Kaiser, D.; Kluge, R.; Neschen, S.; Joost, H.G.; Schürmann, A. Dissociation of Lipotoxicity and Glucotoxicity in a Mouse Model of Obesity Associated Diabetes: Role of Forkhead Box O1 (FOXO1) in Glucose-Induced Beta Cell Failure. *Diabetologia* **2011**, *54*, 605–616. [CrossRef] [PubMed]

44. Honors, M.A.; Davenport, B.M.; Kinzig, K.P. Effects of Consuming a High Carbohydrate Diet after Eight Weeks of Exposure to a Ketogenic Diet. *Nutr. Metab.* **2009**, *6*, 1–9. [CrossRef] [PubMed]

45. Biden, T.J.; Taylor, K.W. Effects of Ketone Bodies on Insulin Release and Islet-Cell Metabolism in the Rat. *Biochem. J.* **1983**, *212*, 371–377. [CrossRef]

46. Rhodes, C.J.; Campbell, I.L.; Szopa, T.M.; Biden, T.J.; Reynolds, P.D.; Fernando, O.N.; Taylor, K.W. Effects of Glucose and D-3-Hydroxybutyrate on Human Pancreatic Islet Cell Function. *Clin. Sci.* **1985**, *68*, 567–572. [CrossRef]

47. Malaisse, W.J.; Lebrun, P.; Yaylali, B.; Camara, J.; Valverde, I.; Sener, A. Ketone Bodies and Islet Function: 45Ca Handling, Insulin Synthesis, and Release. *Am. J. Physiol. Endocrinol. Metab.* **1990**, *259*, E117–E122. [CrossRef]

48. MacDonald, M.J.; Longacre, M.J.; Stoker, S.W.; Brown, L.J.; Hasan, N.M.; Kendrick, M.A. Acetoacetate and β-Hydroxybutyrate in Combination with Other Metabolites Release Insulin from INS-1 Cells and Provide Clues about Pathways in Insulin Secretion. *Am. J. Physiol. Cell Physiol.* **2008**, *294*, C442–C450. [CrossRef]

49. Robinson, A.M.; Williamson, D.H. Physiological Roles of Ketone Bodies as Substrates and Signals in Mammalian Tissues. *Physiol. Rev.* **1980**, *60*, 143–187. [CrossRef]

50. Zhou, Y.-P.; Grill, V. Long Term Exposure to Fatty Acids and Ketones Inhibits B-Cell Functions in Human Pancreatic Islets of Langerhans. *J. Clin. Endocrinol. Metab.* **1995**, *80*, 1584–1590.

51. Pujol, J.B.; Christinat, N.; Ratinaud, Y.; Savoia, C.; Mitchell, S.E.; Dioum, E.H.M. Coordination of GPR40 and Ketogenesis Signaling by Medium Chain Fatty Acids Regulates Beta Cell Function. *Nutrients* **2018**, *10*, 473. [CrossRef]

52. Berne, C. The Effect of Fatty Acids and Ketone Bodies on the Biosynthesis of Insulin in Isolated Pancreatic Islets of Obese Hyperglycemic Mice. *Horm. Metab. Res.* **1975**, *7*, 385–389. [CrossRef] [PubMed]

53. Zhao, C.; Wilson, M.C.; Schuit, F.; Halestrap, A.P.; Rutter, G.A. Expression and Distribution of Lactate/Monocarboxylate Transporter Isoforms in Pancreatic Islets and the Exocrine Pancreas. *Diabetes* **2001**, *50*, 361–366. [CrossRef] [PubMed]

54. Hasan, N.M.; Longacre, M.J.; Seed Ahmed, M.; Kendrick, M.A.; Gu, H.; Ostenson, C.G.; Fukao, T.; MacDonald, M.J. Lower Succinyl-CoA:3-Ketoacid-CoA Transferase (SCOT) and ATP Citrate Lyase in Pancreatic Islets of a Rat Model of Type 2 Diabetes: Knockdown of SCOT Inhibits Insulin Release in Rat Insulinoma Cells. *Arch. Biochem. Biophys.* **2010**, *499*, 62–68. [CrossRef] [PubMed]

55. Berne, C. Determination of D 3 Hydroxybutyrate Dehydrogenase in Mouse Pancreatic Islets with a Photokinetic Technique Using Bacterial Luciferase. *Enzyme* **1976**, *21*, 127–136. [CrossRef] [PubMed]

56. Berne, C. The Metabolism of Lipids in Mouse Pancreatic Islets. The Oxidation of Fatty Acids and Ketone Bodies. *Biochem. J.* **1975**, *152*, 661–666. [CrossRef] [PubMed]

57. Kennedy, A.R.; Pissios, P.; Otu, H.; Xue, B.; Asakura, K.; Furukawa, N.; Marino, F.E.; Liu, F.F.; Kahn, B.B.; Libermann, T.A.; et al. A High-Fat, Ketogenic Diet Induces a Unique Metabolic State in Mice. *Am. J. Physiol. Endocrinol. Metab.* **2007**, *292*, 1724–1739. [CrossRef]

58. Grandl, G.; Straub, L.; Rudigier, C.; Arnold, M.; Wueest, S.; Konrad, D.; Wolfrum, C. Short-Term Feeding of a Ketogenic Diet Induces More Severe Hepatic Insulin Resistance than an Obesogenic High-Fat Diet. *J. Physiol.* **2018**, *596*, 4597–4609. [CrossRef]

59. Jornayvaz, F.R.; Jurczak, M.J.; Lee, H.Y.; Birkenfeld, A.L.; Frederick, D.W.; Zhang, D.; Zhang, X.M.; Samuel, V.T.; Shulman, G.I. A High-Fat, Ketogenic Diet Causes Hepatic Insulin Resistance in Mice, despite Increasing Energy Expenditure and Preventing Weight Gain. *Am. J. Physiol. Endocrinol. Metab.* **2010**, *299*, 808–815. [CrossRef]

60. Morrison, C.D.; Hill, C.M.; DuVall, M.A.; Coulter, C.E.; Gosey, J.L.; Herrera, M.J.; Maisano, L.E.; Sikaffy, H.X.; McDougal, D.H. Consuming a Ketogenic Diet Leads to Altered Hypoglycemiccounter-Regulation in Mice. *J. Diabetes Complicat.* **2020**, *34*, 107557. [CrossRef]

61. Badman, M.K.; Kennedy, A.R.; Adams, A.C.; Pissios, P.; Maratos-Flier, E. A Very Low Carbohydrate Ketogenic Diet Improves Glucose Tolerance in Ob/Ob Mice Independently of Weight Loss. *Am. J. Physiol. Endocrinol. Metab.* **2009**, *297*, E1197–E1204. [CrossRef]

62. Garbow, J.R.; Doherty, J.M.; Schugar, R.C.; Travers, S.; Weber, M.L.; Wentz, A.E.; Ezenwajiaku, N.; Cotter, D.G.; Brunt, E.M.; Crawford, P.A. Hepatic Steatosis, Inflammation, and ER Stress in Mice Maintained Long Term on a Very Low-Carbohydrate Ketogenic Diet. *Am. J. Physiol. Gastrointest. Liver Physiol.* **2011**, *300*, 956–967. [CrossRef] [PubMed]

63. Morens, C.; Sirot, V.; Scheurink, A.J.W.; Van Dijk, G. Low-Carbohydrate Diets Affect Energy Balance and Fuel Homeostasis Differentially in Lean and Obese Rats. *Am. J. Physiol. Regul. Integr. Comp. Physiol.* **2006**, *291*, 1622–1629. [CrossRef] [PubMed]

64. Kinzig, K.P.; Honors, M.A.; Hargrave, S.L. Insulin Sensitivity and Glucose Tolerance Are Altered by Maintenance on a Ketogenic Diet. *Endocrinology* **2010**, *151*, 3105–3114. [CrossRef] [PubMed]

65. Fabbrini, E.; Higgins, P.B.; Magkos, F.; Bastarrachea, R.A.; Saroja Voruganti, V.; Comuzzie, A.G.; Shade, R.E.; Gastaldelli, A.; Horton, J.D.; Omodei, D.; et al. Metabolic Response to High-Carbohydrate and Low-Carbohydrate Meals in a Nonhuman Primate Model. *Am. J. Physiol. Endocrinol. Metab.* **2013**, *304*, 444–451. [CrossRef]

66. Park, S.; Kim, D.S.; Kang, S.; Daily, J.W. A Ketogenic Diet Impairs Energy and Glucose Homeostasis by the Attenuation of Hypothalamic Leptin Signaling and Hepatic Insulin Signaling in a Rat Model of Non-Obese Type 2 Diabetes. *Exp. Biol. Med.* **2011**, *236*, 194–204. [CrossRef]

67. Zhang, X.; Qin, J.; Zhao, Y.; Shi, J.; Lan, R.; Gan, Y.; Ren, H.; Zhu, B.; Qian, M.; Du, B. Long-Term Ketogenic Diet Contributes to Glycemic Control but Promotes Lipid Accumulation and Hepatic Steatosis in Type 2 Diabetic Mice. *Nutr. Res.* **2016**, *36*, 349–358. [CrossRef]

68. Okuda, T.; Morita, N. A Very Low Carbohydrate Ketogenic Diet Increases Hepatic Glycosphingolipids Related to Regulation of Insulin Signalling. *J. Funct. Foods* **2016**, *21*, 70–74. [CrossRef]

69. Westman, E.C.; Yancy, W.S.; Mavropoulos, J.C.; Marquart, M.; McDuffie, J.R. The Effect of a Low-Carbohydrate, Ketogenic Diet versus a Low-Glycemic Index Diet on Glycemic Control in Type 2 Diabetes Mellitus. *Nutr. Metab.* **2008**, *5*, 1–9. [CrossRef]

70. Hall, K.D.; Chen, K.Y.; Guo, J.; Lam, Y.Y.; Leibel, R.L.; Mayer, L.E.S.; Reitman, M.L.; Rosenbaum, M.; Smith, S.R.; Walsh, B.T.; et al. Energy Expenditure and Body Composition Changes after an Isocaloric Ketogenic Diet in Overweight and Obese Men. *Am. J. Clin. Nutr.* **2016**, *104*, 324–333. [CrossRef]

71. Boden, G.; Sargrad, K.; Homko, C.; Mozzoli, M.; Stein, T.P. Effect of a Low-Carbohydrate Diet on Appetite, Blood Glucose Levels, and Insulin Resistance in Obese Patients with Type 2 Diabetes. *Ann. Intern. Med.* **2005**, *142*, 403–411. [CrossRef]

72. Gannon, M.C.; Nuttall, F.Q. Effect of a High-Protein, Low-Carbohydrate Diet on Blood Glucose Control in People with Type 2 Diabetes. *Diabetes* **2004**, *53*, 2375–2382. [CrossRef] [PubMed]

73. Goss, A.M.; Gower, B.A.; Soleymani, T.; Stewart, M.C.; Pendergrass, M.; Lockhart, M.; Krantz, O.; Dowla, S.; Bush, N.; Garr Barry, V.; et al. Effects of Weight Loss during a Very Low Carbohydrate Diet on Specific Adipose Tissue Depots and Insulin Sensitivity in Older Adults with Obesity: A Randomized Clinical Trial. *Metabolism* **2020**, *17*, 1–12. [CrossRef] [PubMed]

74. Rosenbaum, M.; Hall, K.D.; Guo, J.; Ravussin, E.; Mayer, L.S.; Reitman, M.L.; Smith, S.R.; Walsh, B.T.; Leibel, R.L. Glucose and Lipid Homeostasis and Inflammation in Humans Following an Isocaloric Ketogenic Diet. *Obesity* **2019**, *27*, 971–981. [CrossRef] [PubMed]

75. Samaha, F.F.; Iqbal, N.; Seshadri, P.; Chicano, K.L.; Daily, D.A.; McGrory, J.; Williams, T.; Williams, M.; Gracely, E.J.; Stern, L. A Low-Carbohydrate as Compared with a Low-Fat Diet in Severe Obesity. *N. Engl. J. Med.* **2003**, *348*, 2074–2081. [CrossRef]

76. Tay, J.; Luscombe-Marsh, N.D.; Thompson, C.H.; Noakes, M.; Buckley, J.D.; Wittert, G.A.; Yancy Jr, W.S.; Brinkworth, G.D. Comparison of Low- and High-Carbohydrate Diets for Type 2 Diabetes Management: A Randomized Trial. *Am. J. Clin. Nutr.* **2015**, *102*, 780–790. [CrossRef]

77. Allick, G.; Bisschop, P.H.; Ackermans, M.T.; Endert, E.; Meijer, A.J.; Kuipers, F.; Sauerwein, H.P.; Romijn, J.A. A Low-Carbohydrate/High-Fat Diet Improves Glucoregulation in Type 2 Diabetes Mellitus by Reducing Postabsorptive Glycogenolysis. *J. Clin. Endocrinol. Metab.* **2004**, *89*, 6193–6197. [CrossRef]

78. Vernon, M.C.; Mavropoulos, J.; Transue, M.; Yancy, W.S.; Al, V.E.T. Clinical Experience of a Carbohydrate-Restricted Diet: Effect on Diabetes Mellitus. *Metab. Syndr. Relat. Disord.* **2003**, *1*, 233–237. [CrossRef]

79. Hussain, T.A.; Mathew, T.C.; Dashti, A.A.; Asfar, S.; Al-Zaid, N.; Dashti, H.M. Effect of Low-Calorie versus Low-Carbohydrate Ketogenic Diet in Type 2 Diabetes. *Nutrition* **2012**, *28*, 1016–1021. [CrossRef]

80. Yancy, W.S.; Foy, M.; Chalecki, A.M.; Vernon, M.C.; Westman, E.C. A Low-Carbohydrate, Ketogenic Diet to Treat Type 2 Diabetes. *Nutr. Metab.* **2005**, *2*, 34. [CrossRef]

81. Nielsen, J.V.; Joensson, E.A. Low-Carbohydrate Diet in Type 2 Diabetes: Stable Improvement of Bodyweight and Glycemic Control during 44 Months Follow-Up. *Nutr. Metab.* **2008**, *5*, 14. [CrossRef]

82. Müller, J.E.; Sträter-Müller, D.; Marks, H.J.; Gläsner, M.; Kneppe, P.; Clemens-Harmening, B.; Menker, H. Carbohydrate Restricted Diet in Conjunction with Metformin and Liraglutide Is an Effective Treatment in Patients with Deteriorated Type 2 Diabetes Mellitus: Proof-of-Concept Study. *Nutr. Metab.* **2011**, *8*, 92. [CrossRef] [PubMed]

83. Verkoelen, H.; Govers, E.; Maas, H.; Gh, K. Low Carbohydrate Lifestyle Reduces Significantly Insulin Need in Type 2 Diabetes Patients. *Interv. Obes. Diabetes* **2020**, *4*, 411–424.

84. Chen, C.; Id, W.H.; Chen, H.; Chang, C.; Lee, T.; Chen, H.; Kang, Y.; Chie, W.; Id, C.J.; Wang, D.; et al. Effect of a 90 g/Day Low-Carbohydrate Diet on Glycaemic Control, Small, Dense Low-Density Lipoprotein and Carotid Intima-Media Thickness in Type 2 Diabetic Patients: An 18-Month Randomised Controlled Trial. *PLoS ONE* **2020**, *15*, e0240158. [CrossRef] [PubMed]

85. Walton, C.M.; Perry, K.; Hart, R.H.; Berry, S.L.; Bikman, B.T. Improvement in Glycemic and Lipid Profiles in Type 2 Diabetics with a 90-Day Ketogenic Diet. *J. Diabetes Res.* **2019**, *2019*. [CrossRef]

86. McKenzie, A.L.; Hallberg, S.J.; Creighton, B.C.; Volk, B.M.; Link, T.M.; Abner, M.K.; Glon, R.M.; McCarter, J.P.; Volek, J.S.; Phinney, S.D. A Novel Intervention Including Individualized Nutritional Recommendations Reduces Hemoglobin A1c Level, Medication Use, and Weight in Type 2 Diabetes. *JMIR Diabetes* **2017**, *2*, e5. [CrossRef]

87. o'Neill, D.F.; Westman, E.C.; Bernstein, R.K. The Effects of a Low-Carbohydrate Regimen on Glycemic Control and Serum Lipids in Diabetes Mellitus. *Metab. Syndr. Relat. Disord.* **2003**, *1*, 291–298. [CrossRef]

88. Athinarayanan, S.J.; Adams, R.N.; Hallberg, S.J.; McKenzie, A.L.; Bhanpuri, N.H.; Campbell, W.W.; Volek, J.S.; Phinney, S.D.; McCarter, J.P. Long-Term Effects of a Novel Continuous Remote Care Intervention Including Nutritional Ketosis for the Management of Type 2 Diabetes: A 2-Year Nonrandomized Clinical Trial. *Front. Endocrinol.* **2019**, *10*, 348. [CrossRef]

89. Hallberg, S.J.; McKenzie, A.L.; Williams, P.T.; Bhanpuri, N.H.; Peters, A.L.; Campbell, W.W.; Hazbun, T.L.; Volk, B.M.; McCarter, J.P.; Phinney, S.D.; et al. Effectiveness and Safety of a Novel Care Model for the Management of Type 2 Diabetes at 1 Year: An Open-Label, Non-Randomized, Controlled Study. *Diabetes Ther.* **2018**, *9*, 583–612. [CrossRef]

90. Daly, M.E.; Paisey, R.; Paisey, R.; Millward, B.A.; Eccles, C.; Williams, K.; Hammersley, S.; MacLeod, K.M.; Gale, T.J. Short-Term Effects of Severe Dietary Carbohydrate-Restriction Advice in Type 2 Diabetes-A Randomized Controlled Trial. *Diabet. Med.* **2006**, *23*, 15–20. [CrossRef]

91. Muniyappa, R.; Lee, S.; Chen, H.; Quon, M.J. Current Approaches for Assessing Insulin Sensitivity and Resistance in Vivo: Advantages, Limitations, and Appropriate Usage. *Am. J. Physiol. Endocrinol. Metab.* **2008**, *294*, E15–E26. [CrossRef]

92. Lee, S.; Muniyappa, R.; Yan, X.; Chen, H.; Yue, L.Q.; Hong, E.G.; Kim, J.K.; Quon, M.J. Comparison between Surrogate Indexes of Insulin Sensitivity and Resistance and Hyperinsulinemic Euglycemic Clamp Estimates in Mice. *Am. J. Physiol. Endocrinol. Metab.* **2008**, *294*, E261–E270. [CrossRef] [PubMed]

93. Muniyappa, R.; Chen, H.; Muzumdar, R.H.; Einstein, F.H.; Yan, X.; Yue, L.Q.; Barzilai, N.; Quon, M.J. Comparison between Surrogate Indexes of Insulin Sensitivity/Resistance and Hyperinsulinemic Euglycemic Clamp Estimates in Rats. *Am. J. Physiol. Endocrinol. Metab.* **2009**, *297*, E1023–E1029. [CrossRef] [PubMed]

94. Roberts, M.N.; Wallace, M.A.; Tomilov, A.A.; Zhou, Z.; Marcotte, G.R.; Tran, D.; Perez, G.; Gutierrez-Casado, E.; Koike, S.; Knotts, T.A.; et al. A Ketogenic Diet Extends Longevity and Healthspan in Adult Mice. *Cell Metab.* **2017**, *26*, 539–546.e5. [CrossRef] [PubMed]

95. Ribeiro, L.C.; Chittó, A.L.; Müller, A.P.; Rocha, J.K.; Da Silva, M.C.; Quincozes-Santos, A.; Nardin, P.; Rotta, L.N.; Ziegler, D.R.; Gonçalves, C.A.; et al. Ketogenic Diet-Fed Rats Have Increased Fat Mass and Phosphoenolpyruvate Carboxykinase Activity. *Mol. Nutr. Food Res.* **2008**, *52*, 1365–1371. [CrossRef]

96. Hu, S.; Togo, J.; Wang, L.; Wu, Y.; Yang, D.; Xu, Y.; Li, L.; Li, B.; Li, M.; Li, J.; et al. Effects of Dietary Macronutrients and Body Composition on Glucose Homeostasis in Mice. *Natl. Sci. Rev.* **2020**. [CrossRef]

97. Bisschop, P.H.; De Metz, J.; Ackermans, M.T.; Endert, E.; Pijl, H.; Kuipers, F.; Meijer, A.J.; Sauerwein, H.P.; Romijn, J.A. Dietary Fat Content Alters Insulin-Mediated Glucose Metabolism in Healthy Men. *Am. J. Clin. Nutr.* **2001**, *73*, 554–559. [CrossRef]

98. Mulvihill, E.E.; Varin, E.M.; Gladanac, B.; Campbell, J.E.; Ussher, J.R.; Baggio, L.L.; Yusta, B.; Ayala, J.; Burmeister, M.A.; Matthews, D.; et al. Cellular Sites and Mechanisms Linking Reduction of Dipeptidyl Peptidase-4 Activity to Control of Incretin Hormone Action and Glucose Homeostasis. *Cell Metab.* **2017**, *25*, 152–165. [CrossRef]

99. Mulvihill, E.E. Dipeptidyl Peptidase Inhibitor Therapy in Type 2 Diabetes: Control of the Incretin Axis and Regulation of Postprandial Glucose and Lipid Metabolism. *Peptides* **2018**, *100*, 158–164. [CrossRef]

100. Campbell, J.E.; Drucker, D.J. Pharmacology, Physiology, and Mechanisms of Incretin Hormone Action. *Cell Metab.* **2013**, *17*, 819–837. [CrossRef]

101. Stubbs, B.J.; Cox, P.J.; Evans, R.D.; Cyranka, M.; Clarke, K.; De Wet, H. A Ketone Ester Drink Lowers Human Ghrelin and Appetite. *Obesity* **2018**, *26*, 269–273. [CrossRef]

102. Wallenius, V.; Elias, E.; Elebring, E.; Haisma, B.; Casselbrant, A.; Larraufie, P.; Spak, E.; Reimann, F.; Le Roux, C.W.; Docherty, N.G.; et al. Suppression of Enteroendocrine Cell Glucagon-like Peptide (GLP)-1 Release by Fat-Induced Small Intestinal Ketogenesis: A Mechanism Targeted by Roux-En-Y Gastric Bypass Surgery but Not by Preoperative Very-Low-Calorie Diet. *Gut* **2020**, *69*, 1423–1431. [CrossRef] [PubMed]

103. Hirasawa, A.; Tsumaya, K.; Awaji, T.; Katsuma, S.; Adachi, T.; Yamada, M.; Sugimoto, Y.; Miyazaki, S.; Tsujimoto, G. Free Fatty Acids Regulate Gut Incretin Glucagon-like Peptide-1 Secretion through GPR120. *Nat. Med.* **2005**, *11*, 90–94. [CrossRef] [PubMed]

104. Hara, T.; Kashihara, D.; Ichimura, A.; Kimura, I.; Tsujimoto, G.; Hirasawa, A. Role of Free Fatty Acid Receptors in the Regulation of Energy Metabolism. *Biochim. Biophys. Acta Mol. Cell Biol. Lipids* **2014**, *1841*, 1292–1300. [CrossRef] [PubMed]

105. Edfalk, S.; Steneberg, P.; Edlund, H. Gpr40 Is Expressed in Enteroendocrine Cells and Mediates Free Fatty Acid Stimulation of Incretin Secretion. *Diabetes* **2008**, *57*, 2280–2287. [CrossRef]

106. Supale, S.; Li, N.; Brun, T.; Maechler, P. Mitochondrial Dysfunction in Pancreatic β Cells. *Trends Endocrinol. Metab.* **2012**, *23*, 477–487. [CrossRef]

107. Rutter, G.A.; Georgiadou, E.; Martinez-Sanchez, A.; Pullen, T.J. Metabolic and Functional Specialisations of the Pancreatic Beta Cell: Gene Disallowance, Mitochondrial Metabolism and Intercellular Connectivity. *Diabetologia* **2020**, *63*, 1990–1998. [CrossRef]

108. Hu, S.; Wang, L.; Yang, D.; Li, L.; Togo, J.; Wu, Y.; Liu, Q.; Li, B.; Li, M.; Wang, G.; et al. Dietary Fat, but Not Protein or Carbohydrate, Regulates Energy Intake and Causes Adiposity in Mice. *Cell Metab.* **2018**, *28*, 415–431.e4. [CrossRef]

109. Colberg, S.R.; Sigal, R.J.; Fernhall, B.; Regensteiner, J.G.; Blissmer, B.J.; Rubin, R.R.; Chasan-Taber, L.; Albright, A.L.; Braun, B. Exercise and Type 2 Diabetes: The American College of Sports Medicine and the American Diabetes Association: Joint Position Statement. *Diabetes Care* **2010**, *33*, e147–e167. [CrossRef]

110. Monda, V.; Sessa, F.; Ruberto, M.; Carotenuto, M.; Marsala, G.; Monda, M.; Cambria, M.T.; Astuto, M.; Distefano, A.; Messina, G. Aerobic Exercise and Metabolic Syndrome: The Role of Sympathetic Activity and the Redox System. *Diabetes Metab. Syndr. Obes. Targets Ther.* **2020**, *13*, 2433–2442. [CrossRef]

111. Polito, R.; Francavilla, V.C.; Ambrosi, A.; Tartaglia, N.; Tafuri, D.; Monda, M.; Messina, A.; Sessa, F.; Di Maio, G.; Ametta, A.; et al. The Orexin-A Serum Levels Are Strongly Modulated by Physical Activity Intervention in Diabetes Mellitus Patients. *J. Hum. Sport Exerc.* **2020**, *15*, S244–S251.

112. Wroble, K.A.; Trott, M.N.; Schweitzer, G.G.; Rahman, R.S.; Kelly, P.V.; Weiss, E.P. Low-Carbohydrate, Ketogenic Diet Impairs Anaerobic Exercise Performance in Exercise-Trained Women and Men: A Randomized-Sequence Crossover Trial. *J. Sports Med. Phys. Fit.* **2019**, *59*, 600–607. [CrossRef] [PubMed]

113. Pilis, K.; Pilis, A.; Stec, K.; Pilis, W.; Langfort, J.; Letkiewicz, S.; Michalski, C.; Czuba, M.; Zych, M.; Chalimoniuk, M. Three-Year Chronic Consumption of Low-Carbohydrate Diet Impairs Exercise Performance and Has a Small Unfavorable Effect on Lipid Profile in Middle-Aged Men. *Nutrients* **2018**, *10*, 1914. [CrossRef] [PubMed]

114. Ferreira, G.A.; Felippe, L.C.; Silva, R.L.S.; Bertuzzi, R.; De Oliveira, F.R.; Pires, F.O.; Lima-Silva, A.E. Effect of Pre-Exercise Carbohydrate Availability on Fat Oxidation and Energy Expenditure after a High-Intensity Exercise. *Brazilian J. Med. Biol. Res.* **2018**, *51*, 1–8. [CrossRef] [PubMed]

115. Burke, L.M.; Ross, M.L.; Garvican-Lewis, L.A.; Welvaert, M.; Heikura, I.A.; Forbes, S.G.; Mirtschin, J.G.; Cato, L.E.; Strobel, N.; Sharma, A.P.; et al. Low Carbohydrate, High Fat Diet Impairs Exercise Economy and Negates the Performance Benefit from Intensified Training in Elite Race Walkers. *J. Physiol.* **2017**, *595*, 2785–2807. [CrossRef]

116. Greene, D.A.; Varley, B.J.; Hartwig, T.B.; Chapman, P.; Rigney, M. A Low-Carbohydrate Ketogenic Diet Reduces Body Mass without Compromising Performance in Powerlifting and Olympic Weightlifting Athletes. *J. Strength Cond. Res.* **2018**, *32*, 3373–3382. [CrossRef]

117. Paoli, A.; Grimaldi, K.; D'Agostino, D.; Cenci, L.; Moro, T.; Bianco, A.; Palma, A. Ketogenic Diet Does Not Affect Strength Performance in Elite Artistic Gymnasts. *J. Int. Soc. Sports Nutr.* **2012**, *9*, 34. [CrossRef]

118. McSwiney, F.T.; Wardrop, B.; Hyde, P.N.; Lafountain, R.A.; Volek, J.S.; Doyle, L. Keto-Adaptation Enhances Exercise Performance and Body Composition Responses to Training in Endurance Athletes. *Metabolism* **2018**, *81*, 25–34. [CrossRef]

119. Golden, S.H.; Brown, A.; Cauley, J.A.; Chin, M.H.; Gary-Webb, T.L.; Kim, C.; Sosa, J.A.; Sumner, A.E.; Anton, B. Health Disparities in Endocrine Disorders: Biological, Clinical, and Nonclinical Factors–An Endocrine Society Scientific Statement. *J. Clin. Endocrinol. Metab.* **2012**, *97*, E1579–E1639. [CrossRef]

120. O'Neill, B.; Raggi, P. The Ketogenic Diet: Pros and Cons. *Atherosclerosis* **2020**, *292*, 119–126. [CrossRef]

121. Kirkpatrick, C.F.; Bolick, J.P.; Kris-Etherton, P.M.; Sikand, G.; Aspry, K.E.; Soffer, D.E.; Willard, K.E.; Maki, K.C. Review of Current Evidence and Clinical Recommendations on the Effects of Low-Carbohydrate and Very-Low-Carbohydrate (Including Ketogenic) Diets for the Management of Body Weight and Other Cardiometabolic Risk Factors: A Scientific Statement from the Nati. *J. Clin. Lipidol.* **2019**, *13*, 689–711. [CrossRef]

122. Fung, T.T.; Van Dam, R.M.; Hankinson, S.E.; Stampfer, M.; Willett, W.C.; Hu, F.B. Low-Carbohydrate Diets and All-Cause and Cause-Specific Mortality: Two Cohort Studies. *Ann. Intern. Med.* **2010**, *153*, 289–298. [CrossRef] [PubMed]

123. Stern, L.; Iqbal, N.; Seshadri, P.; Chicano, K.L.; Daily, D.A.; McGrory, J.; Williams, M.; Gracely, E.J.; Samaha, F.F. The Effects of Low-Carbohydrate versus Conventional Weight Loss Diets in Severely Obese Adults: One-Year Follow-up of a Randomized Trial. *Ann. Intern. Med.* **2004**, *140*, 778–785. [CrossRef] [PubMed]

124. Tzur, A.; Nijholt, R.; Sparagna, V.; Ritson, A. Adhering to the Ketogenic Diet—Is it Easy or Hard? (Research Review). Available online: https://sci-fit.net/adhere-ketogenic-diet/ (accessed on 18 September 2020).

Publisher's Note: MDPI stays neutral with regard to jurisdictional claims in published maps and institutional affiliations.

 *metabolites* 

Review

Beta Cell Physiological Dynamics and Dysfunctional Transitions in Response to Islet Inflammation in Obesity and Diabetes

Marlon E. Cerf [1,2,3]

1   Grants, Innovation and Product Development, South African Medical Research Council, Tygerberg 7505, South Africa; marlon.cerf@mrc.ac.za
2   Biomedical Research and Innovation Platform, South African Medical Research Council, Tygerberg 7505, South Africa
3   Division of Medical Physiology, Department of Biomedical Sciences, Faculty of Medicine and Health Sciences, University of Stellenbosch, Tygerberg 7505, South Africa

Received: 14 August 2020; Accepted: 10 October 2020; Published: 10 November 2020

Abstract: Beta cells adapt their function to respond to fluctuating glucose concentrations and variable insulin demand. The highly specialized beta cells have well-established endoplasmic reticulum to handle their high metabolic load for insulin biosynthesis and secretion. Beta cell endoplasmic reticulum therefore recognize and remove misfolded proteins thereby limiting their accumulation. Beta cells function optimally when they sense glucose and, in response, biosynthesize and secrete sufficient insulin. Overnutrition drives the pathogenesis of obesity and diabetes, with adverse effects on beta cells. The interleukin signaling system maintains beta cell physiology and plays a role in beta cell inflammation. In pre-diabetes and compromised metabolic states such as obesity, insulin resistance, and glucose intolerance, beta cells biosynthesize and secrete more insulin, i.e., hyperfunction. Obesity is entwined with inflammation, characterized by compensatory hyperinsulinemia, for a defined period, to normalize glycemia. However, with chronic hyperglycemia and diabetes, there is a perpetual high demand for insulin, and beta cells become exhausted resulting in insufficient insulin biosynthesis and secretion, i.e., they hypofunction in response to elevated glycemia. Therefore, beta cell hyperfunction progresses to hypofunction, and may progressively worsen towards failure. Preserving beta cell physiology, through healthy nutrition and lifestyles, and therapies that are aligned with beta cell functional transitions, is key for diabetes prevention and management.

Keywords: beta cell hypofunction; beta cell hyperfunction; ER stress; glucolipotoxicity; oxidative stress

---

1. Introduction

Obesity and diabetes are globally pervasive, increasing in prevalence, and occur over the life-course, often presenting earlier in life e.g., in childhood obesity and diabetes. Urbanization, sedentary lifestyles, and unhealthy diets are some drivers of the global high prevalence of obesity and diabetes, with healthy nutrition particularly important for obesity and diabetes prevention and management. In obesity and early diabetes, overnutrition (i.e., hypercaloric overconsumption, excess nutrients, or nutritional overload, such as chronic high-fat diet (HFD)/high saturated fatty acid (SFA) overconsumption), increases the demand for insulin biosynthesis and secretion, leading to the onset of hyperglycemia and insulin resistance. Beta cells are highly specialized to biosynthesize and secrete insulin to maintain glucose homeostasis. Exposure to chronic hyperglycemia (glucotoxicity) and lipids (lipotoxicity) trigger beta cell dysfunction and death [1–3], and in combination, i.e., an excess of glucose and lipids (glucolipotoxicity), synergize rapid and progressive beta cell demise [4,5].

Glucolipotoxicity therefore is the combined deleterious consequences of elevated chronic glucose and SFA (e.g., palmitic acid) concentrations on specific organs (e.g., the pancreas), micro-organs (e.g., islets) and cells (e.g., beta cells) [6].

2. Systemic and Islet Inflammation in Obesity and Diabetes

Overnutrition, particularly with high SF intake, results in fat accretion, weight gain and obesity [7]. After ingestion, adipose tissue stores ~90% of the non-esterified fatty acid (NEFA) load, and thus substantial adipocyte remodeling occurs. This primarily involves adipose tissue hypertrophy with some hyperplasia to meet triglyceride storage requirements [7]. Hypertrophic adipocytes have a blunted response to insulin and progressively become more lipolytic thereby releasing an excess of NEFA [7] which contributes to systemic insulin resistance. Systemic insulin resistance manifests when glucose is not sufficiently cleared from circulation for uptake in organs for metabolism and storage. This initiates and exacerbates hyperglycemia. In obesity, inflammation leads to systemic insulin resistance and decreases in beta cell mass thereby contributing to beta cell death, dysfunction, failure, and ultimately diabetes. Thus, in obesity, systemic insulin resistance is exacerbated due to hypertrophic adipocytes that do not effectively respond to glucose uptake which worsens glucotoxicity (i.e., hyperglycemia exacerbates); and with excess NEFA, especially SFA release by lipolysis, lipotoxicity compounds and synergizes glucolipotoxicity. In overnutrition and obesity, adiposopathy refers to the response of adipose tissue, behavioral changes and environmental factors [8] characterized by a shift to visceral adipose tissue distribution, a pro-inflammatory imbalance, and ectopic fat deposition when storage capacity exceeds the threshold [9] e.g., in the pancreas. Type 2 diabetic patients had elevated pancreatic triglyceride content [10–13]. Furthermore, pancreatic fat accumulation induced insulin resistance [13] and impaired insulin secretion [14]. Therefore, intra-pancreatic lipids that are lipotoxic, coupled to hyperglycemia that prevails in diabetes, which is glucotoxic, synergize as glucolipotoxicity thereby worsening diabetic outcomes.

An increase in pro-inflammatory mediator gene expression supports the elevated production of cytokines, chemokines, and other pro-inflammatory mediators, such as 12-lipoxygenase (12LO) [7]. In adipose tissue, 12LO recruits and activates immune cells, such as macrophages (M1), natural killer cells (NK), T cells (CD4$^+$ and CD8$^+$) and dendritic cells [7]. Immune cells that infiltrate adipose tissue are pro-inflammatory mediators [7]. Thus, adipose tissue insulin resistance manifests. Various adipokines modulate increased leptin and reduced adiponectin production [7]. Hyperleptinemia and leptin resistance are implicated in obesity, inflammation, and impaired beta cell function. These circulating pro-inflammatory mediators induce beta cell dysfunction [7] characterized by impaired glucose-stimulated insulin secretion (GSIS). In beta cells, chronic high NEFA (particularly SFA) concentrations prompt lipotoxicity, endoplasmic reticulum (ER) and oxidative stress thereby leading to mitochondrial dysfunction [7]. Hyperleptinemia concomitant with beta cell leptin resistance likely impairs GSIS [7] with GSIS further impaired by pro-inflammatory mediators and the infiltrating immune cells in adipose tissue and islets [7]. In obesity and glucolipotoxicity, there is immune cell migration, infiltration and amplification in beta cells which induces and exacerbates beta cell inflammation thereby leading to beta cell demise.

In obesity, chronic inflammation primarily involves mononuclear cells (without an acute immunovascular phase) [15] with a typical 2–3-fold non-site specific increase in pro-inflammatory cytokines and chemokines, i.e., not site specific but manifested in various organs e.g., the eye, kidney, heart, liver, adipose tissue and pancreas [15]. Nuclear factor of kappa light polypeptide gene enhancer in B cells (NFκB) regulates NEFA-induced beta cell inflammation and death [16]. In islets, innate immune system activation, characterized by an increase in innate immune cells and pro-inflammatory mediators, impairs beta cell mass and function [17,18]. Obesity and inflammation (i.e., chemokine and cytokine secretion) impair GSIS [19–21]. Obesity changes the cellular fate of islet macrophages, i.e., reprograms islet macrophages, to acquire specialized functions e.g., synapse formation and cell-cell adhesion [21]. These pathways are activated for macrophage-beta cell interactions [21], which to some extent, explains the transference of vesicles containing insulin from beta cells to macrophages, which is greatly augmented in obesity [21].

The islet macrophages comprise the intra-islet CD11c$^+$ cells (associated with obesity) and peri-islet CD11c$^-$ cells, and, in obesity, myeloid lineage cells dominate islet inflammation [22]. As obesity progresses with weight gain with age, in parallel to chronic inflammation characterized by elevated pro-inflammatory cytokines and chemokines, compensatory mechanisms are induced that shift the homeostatic thresholds, which leads to the onset of diabetes [23]. Inflammation triggers an increase in macrophage migration and infiltration in peripheral organs prompting cellular and organ dysfunction; in beta cells, GSIS is impaired [24–26]. Islet and beta cells can initiate islet inflammation through sensing stimuli and secreting cytokines, chemokines and islet amyloid peptide (IAPP) to activate macrophages. Diabetes is a chronic inflammatory disease and typically islet cell inflammation can be induced in response to systemic inflammation and other stimuli. This increases the complexity in the prevention, pathogenesis and treatment of diabetes, and the preceding pathologies viz. obesity, insulin resistance, glucose intolerance, and beta cell death, dysfunction, and failure.

3. Mediators of Islet Inflammation

Interleukin 1 (IL1) signaling is integral for beta cell physiology and inflammation. Islets richly express the IL1 receptor type I (IL1R) and beta cells are sensitive to its ligands, IL1α and IL1β (a major regulator of inflammation), therefore IL1R signaling regulates beta cell health and physiology [27]. In islets, IL1β secretion occurs due to elevated glucose, thereby initiating inflammation by recruiting and activating macrophages to sustain islet inflammation [28]. The activation of the Nod-like receptor family pyrin domain-containing protein 3 (NLRP3) inflammasome, which produces mature IL1β, was demonstrated in the pancreas and responds to various stimuli [29]. The NLRP3 inflammasome is a critical sensor of nutrient overload, which processes pro-IL1β to its active ILβ in metabolic diseases [30]. The deletion of the NLRP3 inflammasome improves beta cell physiology and viability during oxidative stress and hypoxia that may be associated with anti-inflammatory effects, such as attenuated macrophage islet infiltration [31]. Tumor necrosis factor alpha (TNFα), IL6 and C-X-C motif chemokine ligand 1 (CXCL1) may have additive effects on IL1β [30].

IL1β and insulin have a complex relationship. IL1β and insulin promote each other [30] and have potent effects on glucose homeostasis and inflammation, supporting their roles in the physiology and pathology of metabolism [30]. In macrophages, IL1β and insulin increased glucose uptake, and insulin reinforced inflammation through insulin receptor signaling, glucose metabolism, reactive oxygen species (ROS) production, and NLRP3 inflammasome-mediated IL1β secretion [30]. Insulin enhances IL1β production [30], and up-regulates insulin receptors, phosphatidylinositol 3-kinase-protein kinase B (PI3K-Akt) signaling, glucose transporter 1 (GLUT1)-mediated glucose uptake, glucose metabolism, and ROS generation to activate the NLRP3 inflammasome [32,33]. Furthermore, insulin and IL1β stimulate glucose uptake in muscle and fat (for glucose disposal), and in immune cells (to fuel the immune system) [30].

IL antagonism presents an attractive strategy for beta cell protection and preservation. In macrophages and islets, blocking IL1 signaling with an IL1R antagonist (IL1Ra) limited IAPP-induced secretion of pro-inflammatory cytokines [34], suggesting that IL1 was required for islet inflammation in IAPP formation [35]. A cell cycle regulatory factor, p27^{kip1} (p27), plays a role in the proliferation of inflammatory cells as p27$^{-/-}$ mice had severe functional islet injury, with increased circulating IL1 and TNFα levels that induced macrophage proliferation [36]. In beta cells, another cell cycle regulator and transcription factor, E2F1, regulated insulin secretion through the Kir6.2 promoter [37,38]. E2F1 overexpression partially prevented IL1β-mediated inhibition of Kir6.2 expression and on GSIS [39]. In mice with beta cell IL1Ra knockout, the reduction in proliferation genes resulted in impaired beta cell proliferation and function, through the E2F1-Kir6.2 pathway; thus the benefits of IL1 antagonism appear to be related to beta cell turnover and function [39] and may provide insights in the treatment of diabetes.

Beta cell microRNAs (miRs) are also implicated in islet inflammation and associated with IL signaling [40–42]. Pro-inflammatory mediators induced miR-146a-5p expression in human islets, MIN6 mouse beta cells and INS1 rat beta cells [41,43,44], to adapt to IL1β-mediated NFκB activation (as there are NFκB binding elements on the miR-146a promoter [41,45,46]). In INS1 cells transfected with miR-146a-5p, there was reduced NFκB and inducible nitric oxide synthase (iNOS) promotor activity that subsequently decreased cytokine-mediated iNOS protein expression, nitic oxide (NO) synthesis and mitogen-activated protein kinase (MAPK) signaling [44]. Therefore, miR-146a-5p down-regulated islet inflammation and beta cell death by impairing NFκB and MAPK signaling [44].

In the inflamed beta cells of hyperglycemic non-obese diabetic mice, miR-203a (which was up-regulated) targeted insulin receptor substrate 2 (IRS2) thereby regulating beta cell proliferation and apoptosis [47]. In MIN6 cells, IRS2 overexpression may protect from the reduced proliferation and apoptosis induced by miR-203a [47]. Furthermore, MIN6 cells transfected with Irs2 siRNA antagonized the effects of miR203a inhibitors which suggested that Irs2 and IRS2 may be novel targets of miR-203a; and miR-203a inhibitors and IRS2 may be novel therapies [47].

4. Metabolic Interplay of Islet Inflammation, Glucolipotoxicity, and Beta Cell Dysfunction

In diabetes, several factors drive a decline in beta cell mass including cellular stress due to compensatory insulin overproduction [48], glucotoxicity due to chronic hyperglycemia, lipotoxicity due to chronic high dietary SFA consumption [49], glucolipotoxicity, inflammation [32], autoimmunity [50] and obesity [3]. However, there are interrelations which require unravelling. High dietary SFA intake contributes to glucotoxicity. Prolonged SFA exposure induces lipotoxicity that diminishes beta cell mass and function, thereby contributing to and exacerbating hyperglycemia. Persistent lipotoxicity thus exacerbates beta cell glucotoxicity i.e., diminished beta cell mass and function will inadequately respond to elevations in glycemia and hyperglycemia will ensue. Persistent hyperglycemia will further augment glucotoxicity for beta cells. In beta cells, glucotoxicity and resident pro-inflammatory macrophages contribute to increased IL1β levels and subsequently impaired GSIS [28]. In addition, glucolipotoxicity, with its synergistic deleterious effects, will have even worse outcomes for beta cells.

Moreover, obesity is entwined with inflammation and characterized by compensatory hyperinsulinemia. Overburdening the beta cells, e.g., in hyperinsulinemia, results in the emergence of cellular stress evident by ER and oxidative stress. Cellular stress is accompanied by islet inflammation that further exacerbates islet and beta cell demise; in addition, autoimmunity is a trait of compromised beta cells in type 1 diabetes that prompts beta cell death. The obese-inflammatory metabolic state is fueled by aberrant cytokine generation, increased synthesis of acute-phase reactants (such as C-reactive protein (CRP)), and an activated pro-inflammatory response [51]. In adults, latent autoimmune diabetes (LADA) is characterized by mild autoimmunity with gradual progression to insulin dependence (relative to type 1 diabetes), concomitant with some type 2 diabetes features such as insulin resistance [52] and overweight/obesity [53]. These interrelations reflect the complexity in beta cell mechanisms and dynamics as they adapt (i.e., compensate) to maintain function in response to variable metabolic demands.

In murine and human islets, ER stress can also be induced by palmitate [54]. The beta cell can itself activate islet inflammation as resident macrophages sense beta cell activity by reacting to ATP, which is co-secreted with insulin, thereby activating macrophages and resulting in inflammation [55]. Beta cell self-activation of islet inflammation is an interesting phenomenon that requires more investigation into what triggers the process, and further delineation of islet inflammation and autoimmunity. This may provide insights on beta cell dedifferentiation to escape attack and death.

5. Beta Cell Dysfunctional Transitions

5.1. Beta Cell Physiology (Optimal Beta Cell Function)

Beta cells require normal beta cell integrity, i.e., number, size, and machinery, to effectively respond to constantly fluctuating metabolic demand for insulin [56] by rapidly equilibrating glucose across the plasma membrane for insulin exocytosis [57]. This requires adequate insulin biosynthesis and maintaining beta cell readiness for GSIS while regulating glycemia within a tight physiological range; and depends on the coordinated regulation of insulin secretion through nutrient availability, hormonal and neural inputs [58]. As the key regulator of beta cell physiology, glucose coordinates and stimulates insulin gene transcription, pro/insulin biosynthesis and protein translation, and insulin secretion [59] and maintains the highly differentiated and specialized beta cells to meet insulin biosynthesis and secretory demands [60].

Furthermore, beta cells are equipped with highly developed and active ER, for their role in folding, export, and processing of newly biosynthesized insulin [61]. Maintaining optimal beta cell physiology for insulin biosynthesis and secretion requires efficiently functioning unfolded protein response (UPR) machinery [57]. In beta cells, nuclear factor erythroid 2p45-related factor 2 (Nrf2) regulates many antioxidant defense factors [62]. ROS are necessary for beta cell signaling but excessive ROS induces oxidative stress. Physiologically, in response to Nrf2 activators, GSIS was impaired with altered ROS handling [62]. Nrf2 activators may shield beta cells from glucolipotoxicity, through the preservation of mitochondrial function and redox balance, to maintain GSIS [62].

The IL1R signaling system maintains beta cell physiology—with pancreatic IL1R knockout, in lean mice, whole body glucose homeostasis was disrupted which was exacerbated in obese db/db mice, concomitant with reduced insulin content and GSIS, both in vivo and ex vivo [27]. Furthermore, in mouse and human islets ex vivo, IL1R signaling enhanced the docking of insulin granules, supporting a role for IL1R activation in refining beta cell physiology [27,63]. Optimal beta cell physiology is therefore characterized by sufficient insulin biosynthesis and secretion; in the ER, pro/insulin protein misfolding is well within the threshold and therefore proteostasis is maintained; with inflammation in the physiological range (Figure 1). Healthy beta cells maintain their function through adequate beta cell mass, and stable beta cell death, which prevents diabetes by keeping metabolism and physiology intact (Figure 1).

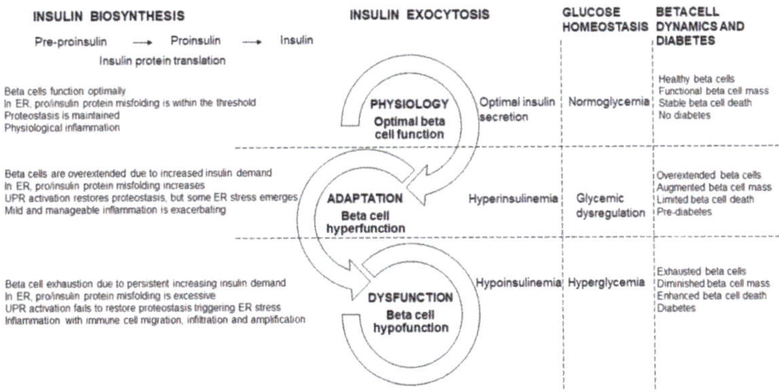

Figure 1. Beta cell physiology and dysfunctional transitions.

5.2. Beta Cell Hyperfunction

Physiologically, healthy beta cells respond to the nutritional cues with biphasic insulin secretion: the first phase needs a fast and substantial increase in intracellular calcium concentrations to release insulin granules from a limited pool, whereas the second phase also requires an increase in intracellular calcium concentrations but is amplified by glucose to replenish the pool of releasable granules [64]. However, in overnutrition, there is excessive nutrient secretagogues e.g., glucose, lipids, and amino acids that initially induces mild nutritional metabolic stress on the beta cells, thereby elevating basal and enhancing amplification of insulin secretion, resulting in hyperinsulinemia [57] that reflects hyperfunction. During pre-diabetes and early diabetes, beta cells adapt in response to muscle, hepatic and adipose tissue insulin resistance by enhancing GSIS [62] and their hyperfunction largely contributes to increased insulin output, which is supported by beta cell mass expansion [65]. Beta cell mass expands by hyperplasia through increased replication and neogenesis from the recruitment of progenitors, and hypertrophy to establish and sustain hyperinsulinemia [66] thereby restoring glycemia. In insulin resistant and impaired glucose tolerant individuals, beta cell mass is primarily replenished through neogenesis [67,68]. Beta cell hyperplasia, concomitant with beta cell hypertrophy, initially support beta cell adaptation to restore glucose homeostasis through increased insulin biosynthesis and secretion to meet greater demand.

Often beta cell hyperfunction has adverse consequences. In db/db mice [69], ZDF rats [24,70], and obese humans [63,71], the up-regulation of the IL1R ligands supports the adaptive response to sustain insulin biosynthesis and secretion [27]. In early diabetes, an increase in insulin may drive and sustain inflammation in macrophages and thereby contribute to the chronic low-grade inflammation [30]. In beta cell hyperfunction, protein misfolding starts to breach the cellular threshold as ER start to overload. Beta cells become overextended due to the progressively increasing insulin demand (Figure 1). Furthermore, in ER, pro/insulin protein misfolding increases but remains within the threshold, and although the UPR activation restores some proteostasis, ER stress starts to emerge (Figure 1). Beta cell destruction by the autoimmune infiltration continues and is exacerbated by the increasing metabolic and glycemic overload causing ER and oxidative stress and apoptosis [72]. Physiologically, the pro-apoptotic factors, Bax and Bak, do not influence glucose-stimulated Ca^{2+} responses, insulin secretion or glucose tolerance [73]. However, unresolved ER stress triggered Bax and Bak-dependent death in murine fibroblasts [74–77]; in islets, during early glucolipotoxicity-induced ER stress, Bax and Bak deletion amplified the UPR [73]. Oxidative stress is also induced during beta cell hyperfunction. The pro-inflammatory response is mild and manageable. Hyperinsulinemia helps to maintain some a level of normoglycemia, through a greater functional load per beta cell, with glycemia steadily increasing towards glucose dysregulation (Figure 1). During hyperfunction in pre-diabetes, the functionally overextended beta cells are supported by enhanced mass, with beta cell death playing a limited role (Figure 1).

5.3. Beta Cell Hypofunction

When the beta cell adaptation fails, hyperglycemia and diabetes develop due to insulin deficiency [65]. The shift from beta cell hyperfunction to hypofunction is driven by increasing severity in hyperglycemia, hyperlipidemia, and glucolipotoxicity. In beta cells, ROS/reactive nitrogen species (RNS) production can exacerbate ER stress and induce cell death [78] that progressively diminishes beta cell function [66]. Islet inflammation is attributed to overnutrition, e.g., glucolipotoxicity, that leads to beta cell exhaustion, cytokine and chemokine synthesis, that triggers immune cells recruitment to islets and beta cells [79] to further amplify beta cell dysfunction and insulin resistance [56]. Increased protein misfolding and ER overloading leading to ER stress, concomitant with oxidative stress (as increases in ROS/RNS are not sufficiently neutralized by antioxidant enzymes) and inflammation all converge towards further deteriorating beta cell dysfunction. Chronic IL1 exposure contributed to beta cell death and dysfunction by altering gene transcription, protein activity and inducing oxidative stress [80]. As diabetes progresses, increased IL1β expression was linked to insulin resistance and beta cell destruction [19]; with damaged beta cells eventually failing to respond to increasing insulin demand [62], due their rapid death and functional impairment [65]. As beta cells are overextended towards exhaustion, they hypofunction characterized by hypoplasia, hypotrophy, reduced proliferation and neogenesis and dedifferentiation [66] as they lose their specialized function and identity. Thus, the highly differentiated and specialized state of beta cells is eroded, and they are reduced in number and size (through beta cell death and non-replenishment from progenitors), which renders a sub-critical mass of residual beta cells to maintain their physiology. Beta cell dedifferentiation also contributes to reduced beta cell mass and a marked reduction in beta cell mass presents closer to overt diabetes [65]. As beta cell dysfunction further deteriorates, beta cells cannot meet the persistent demand for increased insulin biosynthesis and secretion in response to persevering elevated glycemia i.e., chronic hyperglycemia. Prolonged chronic hyperglycemia triggers beta cell exhaustion resulting in hypofunction, characterized by hypoinsulinemia which is insufficiently responsive to increases in glycemia. In overt beta cell dysfunction, when beta cells are exhausted towards failure, there is reduced insulin biosynthesis and insufficient insulin secretion [66]—i.e., more insulin is required, e.g., exogenously to maintain glucose homeostasis—and there is decreased insulin signaling in peripheral organs i.e., reduced glucose uptake, with prevailing hyperglycemia that progressively exacerbates.

Beta cells that are continuously overextended by persistently increased insulin demand become exhausted (Figure 1). In ER, pro/insulin protein misfolding increases beyond the threshold, and with the ER overload, the UPR is activated but does not restore proteostasis, therefore ER stress manifests and is exacerbated by oxidative stress (Figure 1). Islet and beta cell pro-inflammatory responses are triggered with immune cell migration, infiltration and amplification (Figure 1). Inflammatory pathways are a complex, intertwined, cascading sequence of events that converge to induce beta cell death, and diminish beta cell mass [7] and function leading to beta cell failure and diabetes. In diabetes, in inflamed islets, the islet macrophage inflammasome-IL1β pathway is a common pathway for beta cell dysfunction [81]. In rodent diabetes, islet inflammation and macrophage infiltration induce beta cell failure [81]. Chronic islet and beta cell glucolipotoxicity induces and exacerbates beta cell dysfunction and death, and accelerates beta cell metabolic overload, thereby impairing beta cell adaptation that progresses towards beta cell failure [82]. During hypofunction, beta cells become exhausted as increased beta cell death contributes to reduced beta cell mass (Figure 1). Fatigued beta cells fail and diabetes manifests.

5.4. Beta Cell Failure and Diabetes

Beta cell dysfunctional transitions are initiated with hyperfunction that precedes hypofunction and advances to beta cell failure. Beta cell death occurs in healthy beta cells, but with the dysfunctional transitions i.e., from hyper- to hypofunction to failure, the beta cell population progressively diminishes through increased beta cell death. Although beta cells are resilient and will compensate to cope with insulin demand despite reduced numbers [56], with persistent hyperglycemia, there is increased beta cell death, and with less beta cells available to function (i.e., inadequate beta cell capacity), they are therefore overburdened towards exhaustion, beta cell failure manifests, and diabetes ensues.

6. Conclusions

Beta cells serve to prevent metabolic diseases by biosynthesizing and secreting insulin to maintain glucose homeostasis. Stressed and inflamed beta cells are functionally compromised and cannot adapt to effectively respond to increased insulin demand which advances beta cell dysfunction towards failure and diabetes. Beta cell preservation and replenishment, through novel agents, healthy lifestyles of balanced diets and regular exercise, maintains beta cell physiology and sustains adaptation [56]. Novel therapies should focus on beta cell protection and functional recovery in early diabetes, and support beta cell mass replacement in late diabetes [65]. Targeting pro-inflammatory mediators and redox balance are viable strategies for diabetes prevention and treatment, given their roles in the beta cell dysfunctional transitions.

Funding: The author is supported by the National Research Foundation (South Africa).

Conflicts of Interest: The authors declare no conflict of interest.

References

1. Jeffrey, K.D.; Alejandro, E.U.; Luciani, D.S.; Kalynyak, T.B.; Hu, X.; Li, H.; Lin, Y.; Townsend, R.R.; Polonsky, K.S.; Johnson, J.D. Carboxypeptidase E mediates palmitate-induced β-cell ER stress and apoptosis. *Proc. Natl. Acad. Sci. USA* **2008**, *105*, 8452–8457. [CrossRef]

2. Johnson, J.D.; Luciani, D.S. Mechanisms of pancreatic β-cell apoptosis in diabetes and its therapies. *Adv. Exp. Med. Biol.* **2010**, *654*, 447–462. [CrossRef]

3. Butler, A.E.; Janson, J.; Bonner-Weir, S.; Ritzel, R.; Rizza, R.A.; Butler, P.C. β-cell deficit and increased β-cell apoptosis in humans with type 2 diabetes. *Diabetes* **2003**, *52*, 102–110. [CrossRef]

4. Tanabe, K.; Liu, Y.; Hasan, S.D.; Martinez, S.C.; Cras-Méneur, C.; Welling, C.M.; Bernal-Mizrachi, E.; Tanizawa, Y.; Rhodes, C.J.; Zmuda, E.; et al. Glucose and fatty acids synergize to promote β-cell apoptosis through activation of glycogen synthase kinase 3β independent of JNK activation. *PLoS ONE* **2011**, *6*, e18146. [CrossRef] [PubMed]

5. Poitout, V.; Robertson, R.P. Glucolipotoxicity: Fuel excess and β-cell dysfunction. *Endocr. Rev.* **2008**, *29*, 351–366. [CrossRef] [PubMed]

6. Poitout, V.; Amyot, J.; Semache, M.; Zarrouki, B.; Hagman, D.; Fontés, G. Glucolipotoxicity of the pancreatic beta cell. *Biochim. Biophys. Acta Mol. Cell Biol. Lipids* **2010**, *1801*, 289–298. [CrossRef] [PubMed]

7. Imai, Y.; Dobrian, A.D.; Morris, M.A.; Nadler, J.L. Islet inflammation: A unifying target for diabetes treatment? *Trends Endocrinol. Metab.* **2013**, *24*, 351–360. [CrossRef] [PubMed]

8. Bays, H.E. Adiposopathy: Is 'sick fat' a cardiovascular disease? *J. Am. Coll. Cardiol.* **2011**, *57*, 2461–2473. [CrossRef]

9. Carbone, F.; Lattanzio, M.S.; Minetti, S.; Ansaldo, A.M.; Ferrara, D.; Molina-Molina, E.; Belfiore, A.; Elia, E.; Pugliese, S.; Palmieri, V.O.; et al. Circulating CRP levels are associated with epicardial and visceral fat depots in women with metabolic syndrome criteria. *Int. J. Mol. Sci.* **2019**, *20*, 5981. [CrossRef]

10. Macauley, M.; Percival, K.; Thelwall, P.E.; Hollingsworth, K.G.; Taylor, R. Altered volume, morphology and composition of the pancreas in type 2 diabetes. *PLoS ONE* **2015**, *10*, e0126825. [CrossRef]

11. Szczepaniak, L.S.; Victor, R.G.; Mathur, R.; Nelson, M.D.; Szczepaniak, E.W.; Tyer, N.; Chen, I.; Unger, R.H.; Bergman, R.N.; Lingvay, I. Pancreatic steatosis and its relationship to β-cell dysfunction in humans: Racial and ethnic variations. *Diabetes Care* **2012**, *35*, 2377–2383. [CrossRef]

12. Tushuizen, M.E.; Bunck, M.C.; Pouwels, P.J.; Bontemps, S.; Van Waesberghe, J.H.; Schindhelm, R.K.; Mari, A.; Heine, R.J.; Diamant, M. Pancreatic fat content and β-cell function in men with and without type 2 diabetes. *Diabetes Care* **2007**, *30*, 2916–2921. [CrossRef]

13. Van Der Zijl, N.J.; Goossens, G.H.; Moors, C.C.; Van Raalte, D.H.; Muskiet, M.H.; Pouwels, P.J.; Blaak, E.E.; Diamant, M. Ectopic fat storage in the pancreas, liver, and abdominal fat depots: Impact on β-cell function in individuals with impaired glucose metabolism. *J. Clin. Endocrinol. Metab.* **2011**, *96*, 459–467. [CrossRef]

14. Heni, M.; Machann, J.; Staiger, H.; Schwenzer, N.F.; Peter, A.; Schick, F.; Claussen, C.D.; Stefan, N.; Häring, H.U.; Fritsche, A. Pancreatic fat is negatively associated with insulin secretion in individuals with impaired fasting glucose and/or impaired glucose tolerance: A nuclear magnetic resonance study. *Diabetes Metab. Res. Rev.* **2010**, *26*, 200–205. [CrossRef] [PubMed]

15. Böni-Schnetzler, M.; Meier, D.T. Islet inflammation in type 2 diabetes. *Semin. Immunopathol.* **2019**, *41*, 501–513. [CrossRef]

16. Choi, H.J.; Hwang, S.; Lee, S.H.; Lee, Y.R.; Shin, J.; Park, K.S.; Cho, Y.M. Genome-wide identification of palmitate-regulated immediate early genes and target genes in pancreatic beta-cells reveals a central role of NF-κB. *Mol. Biol. Rep.* **2012**, *39*, 6781–6789. [CrossRef]

17. Donath, M.Y.; Dalmas, É.; Sauter, N.S.; Böni-Schnetzler, M. Inflammation in obesity and diabetes: Islet dysfunction and therapeutic opportunity. *Cell Metab.* **2013**, *17*, 860–872. [CrossRef] [PubMed]

18. Donath, M.Y.; Halban, P.A. Decreased beta-cell mass in diabetes: Significance, mechanisms and therapeutic implications. *Diabetologia* **2004**, *47*, 581–589. [CrossRef]

19. Donath, M.Y.; Schumann, D.M.; Faulenbach, M.; Ellingsgaard, H.; Perren, A.; Ehses, J.A. Islet inflammation in type 2 diabetes: From metabolic stress to therapy. *Diabetes Care* **2008**, *31*, S161–S164. [CrossRef]

20. Eguchi, K.; Manabe, I.; Oishi-Tanaka, Y.; Ohsugi, M.; Kono, N.; Ogata, F.; Yagi, N.; Ohto, U.; Kimoto, M.; Miyake, K.; et al. Saturated fatty acid and TLR signaling link β cell dysfunction and islet inflammation. *Cell Metab.* **2012**, *15*, 518–533. [CrossRef]

21. Ying, W.; Lee, Y.S.; Dong, Y.; Seidman, J.S.; Yang, M.; Isaac, R.; Seo, J.B.; Yang, B.H.; Wollam, J.; Riopel, M.; et al. Expansion of islet-resident macrophages leads to inflammation affecting β cell proliferation and function in obesity. *Cell Metab.* **2019**, *29*, 457–474.e5. [CrossRef] [PubMed]

22. Starling, S. Drivers of islet inflammation. *Nat. Rev. Endocrinol.* **2019**, *15*, 128. [CrossRef] [PubMed]

23. Medzhitov, R. Inflammation 2010: New adventures of an old flame. *Cell* **2010**, *140*, 771–776. [CrossRef] [PubMed]

24. Jourdan, T.; Godlewski, G.; Cinar, R.; Bertola, A.; Szanda, G.; Liu, J.; Tam, J.; Han, T.; Mukhopadhyay, B.; Skarulis, M.C.; et al. Activation of the Nlrp3 inflammasome in infiltrating macrophages by endocannabinoids mediates beta cell loss in type 2 diabetes. *Nat. Med.* **2013**, *19*, 1132–1140. [CrossRef] [PubMed]

25. Kahn, S.E.; Hull, R.L.; Utzschneider, K.M. Mechanisms linking obesity to insulin resistance and type 2 diabetes. *Nature* **2006**, *444*, 840–846. [CrossRef]

26. Richardson, S.J.; Willcox, A.; Bone, A.J.; Foulis, A.K.; Morgan, N.G. Islet-associated macrophages in type 2 diabetes. *Diabetologia* **2009**, *52*, 1686–1688. [CrossRef]

27. Burke, S.J.; Batdorf, H.M.; Burk, D.H.; Martin, T.M.; Mendoza, T.; Stadler, K.; Alami, W.; Karlstad, M.D.; Robson, M.J.; Blakely, R.D.; et al. Pancreatic deletion of the interleukin-1 receptor disrupts whole body glucose homeostasis and promotes islet β-cell de-differentiation. *Mol. Metab.* **2018**, *14*, 95–107. [CrossRef]

28. Herder, C.; Dalmas, E.; Böni-Schnetzler, M.; Donath, M.Y. The IL-1 pathway in type 2 diabetes and cardiovascular complications. *Trends Endocrinol. Metab.* **2015**, *26*, 551–563. [CrossRef]

29. Kammoun, H.L.; Allen, T.L.; Henstridge, D.C.; Barre, S.; Coll, R.C.; Lancaster, G.I.; Cron, L.; Reibe, S.; Chan, J.Y.; Bensellam, M.; et al. Evidence against a role for NLRP3-driven islet inflammation in db/db mice. *Mol. Metab.* **2018**, *10*, 66–73. [CrossRef]

30. Dror, E.; Dalmas, E.; Meier, D.T.; Wueest, S.; Thévenet, J.; Thienel, C.; Timper, K.; Nordmann, T.M.; Traub, S.; Schulze, F.; et al. Postprandial macrophage-derived IL-1β stimulates insulin, and both synergistically promote glucose disposal and inflammation. *Nat. Immunol.* **2017**, *18*, 283–292. [CrossRef]

31. Sokolova, M.; Sahraoui, A.; Høyem, M.; Øgaard, J.; Lien, E.; Aukrust, P.; Yndestad, A.; Ranheim, T.; Scholz, H. Nlrp3 inflammasome mediates oxidative stress-induced pancreatic islet dysfunction. *Am. J. Physiol. Endocrinol. Metab.* **2018**, *315*, E912–E923. [CrossRef] [PubMed]

32. Zhou, R.; Tardivel, A.; Thorens, B.; Choi, I.; Tschopp, J. Thioredoxin-interacting protein links oxidative stress to inflammasome activation. *Nat. Immunol.* **2010**, *11*, 136–140. [CrossRef] [PubMed]

33. Pétrilli, V.; Papin, S.; Dostert, C.; Mayor, A.; Martinon, F.; Tschopp, J. Activation of the NALP3 inflammasome is triggered by low intracellular potassium concentration. *Cell Death Differ.* **2007**, *14*, 1583–1589. [CrossRef] [PubMed]

34. Westwell-Roper, C.; Dai, D.L.; Soukhatcheva, G.; Potter, K.J.; Van Rooijen, N.; Ehses, J.A.; Verchere, C.B. IL-1 blockade attenuates islet amyloid polypeptide-induced proinflammatory cytokine release and pancreatic islet graft dysfunction. *J. Immunol.* **2011**, *187*, 2755–2765. [CrossRef] [PubMed]

35. Westwell-Roper, C.Y.; Chehroudi, C.A.; Denroche, H.C.; Courtade, J.A.; Ehses, J.A.; Verchere, C.B. IL-1 mediates amyloid-associated islet dysfunction and inflammation in human islet amyloid polypeptide transgenic mice. *Diabetologia* **2015**, *58*, 575–585. [CrossRef]

36. Li, Y.; Ding, X.; Fan, P.; Guo, J.; Tian, X.; Feng, X.; Zheng, J.; Tian, P.; Ding, C.; Xue, W. Inactivation of p27^{kip1} promoted nonspecific inflammation by enhancing macrophage proliferation in islet transplantation. *Endocrinology* **2016**, *157*, 4121–4132. [CrossRef]

37. Annicotte, J.S.; Blanchet, E.; Chavey, C.; Iankova, I.; Costes, S.; Assou, S.; Teyssier, J.; Dalle, S.; Sardet, C.; Fajas, L. The CDK4-pRB-E2F1 pathway controls insulin secretion. *Nat. Cell Biol.* **2009**, *11*, 1017–1023. [CrossRef]

38. Grouwels, G.; Cai, Y.; Hoebeke, I.; Leuckx, G.; Heremans, Y.; Ziebold, U.; Stangé, G.; Chintinne, M.; Ling, Z.; Pipeleers, D.; et al. Ectopic expression of E2F1 stimulates β-cell proliferation and function. *Diabetes* **2010**, *59*, 1435–1444. [CrossRef]

39. Böni-Schnetzler, M.; Häuselmann, S.P.; Dalmas, E.; Meier, D.T.; Thienel, C.; Traub, S.; Schulze, F.; Steiger, L.; Dror, E.; Martin, P.; et al. β cell-specific deletion of the IL-1 receptor antagonist impairs β cell proliferation and insulin secretion. *Cell Rep.* **2018**, *22*, 1774–1786. [CrossRef]

40. Osmai, M.; Osmai, Y.; Bang-Berthelsen, C.H.; Pallesen, E.M.; Vestergaard, A.L.; Novotny, G.W.; Pociot, F.; Mandrup-Poulsen, T. MicroRNAs as regulators of beta-cell function and dysfunction. *Diabetes Metab. Res.* **2016**, *32*, 334–349. [CrossRef]

41. Roggli, E.; Britan, A.; Gattesco, S.; Lin-Marq, N.; Abderrahmani, A.; Meda, P.; Regazzi, R. Involvement of microRNAs in the cytotoxic effects exerted by proinflammatory cytokines on pancreatic β-cells. *Diabetes* **2010**, *59*, 978–986. [CrossRef] [PubMed]

42. Grieco, F.A.; Sebastiani, G.; Juan-Mateu, J.; Villate, O.; Marroqui, L.; Ladrière, L.; Tugay, K.; Regazzi, R.; Bugliani, M.; Marchetti, P.; et al. MicroRNAs miR-23a-3p, miR-23b-3p, and miR-149-5p regulate the expression of proapoptotic bh3-only proteins DP5 and PUMA in human pancreatic β-cells. *Diabetes* **2017**, *66*, 100–112. [CrossRef] [PubMed]

43. Fred, R.G.; Bang-Berthelsen, C.H.; Mandrup-Poulsen, T.; Grunnet, L.G.; Welsh, N. High glucose suppresses human islet insulin biosynthesis by inducing miR-133a leading to decreased polypyrimidine tract binding protein expression. *PLoS ONE* **2010**, *5*, e10843. [CrossRef] [PubMed]

44. Vestergaard, A.L.; Bang-Berthelsen, C.H.; Fløyel, T.; Stahl, J.L.; Christen, L.; Sotudeh, F.T.; De Hemmer Horskjñr, P.; Frederiksen, K.S.; Kofod, F.G.; Bruun, C.; et al. MicroRNAs and histone deacetylase inhibition-mediated protection against inflammatory β-cell damage. *PLoS ONE* **2018**, *13*, e0203713. [CrossRef]

45. Taganov, K.D.; Boldin, M.P.; Chang, K.J.; Baltimore, D. NF-κB-dependent induction of microRNA miR-146, an inhibitor targeted to signaling proteins of innate immune responses. *Proc. Natl. Acad. Sci. USA* **2006**, *103*, 12481–12486. [CrossRef]

46. Cameron, J.E.; Yin, Q.; Fewell, C.; Lacey, M.; McBride, J.; Wang, X.; Lin, Z.; Schaefer, B.C.; Flemington, E.K. Epstein-Barr virus latent membrane protein 1 induces cellular microRNA miR-146a, a modulator of lymphocyte signaling pathways. *J. Virol.* **2008**, *82*, 1946–1958. [CrossRef]

47. Duan, X.; Zhao, L.; Jin, W.; Xiao, Q.; Peng, Y.; Huang, G.; Li, X.; DaSilva-Arnold, S.; Yu, H.; Zhou, Z. MicroRNA-203a regulates pancreatic β cell proliferation and apoptosis by targeting IRS2. *Mol. Biol. Rep.* **2020**. [CrossRef]

48. Back, S.H.; Kaufman, R.J. Endoplasmic reticulum stress and type 2 diabetes. *Annu. Rev. Biochem.* **2012**, *81*, 767–793. [CrossRef]

49. Laybutt, D.R.; Preston, A.M.; Åkerfeldt, M.C.; Kench, J.G.; Busch, A.K.; Biankin, A.V.; Biden, T.J. Endoplasmic reticulum stress contributes to beta cell apoptosis in type 2 diabetes. *Diabetologia* **2007**, *50*, 752–763. [CrossRef]

50. Cervin, C.; Lyssenko, V.; Bakhtadze, E.; Lindholm, E.; Nilsson, P.; Tuomi, T.; Cilio, C.M.; Groop, L. Genetic similarities between latent autoimmune diabetes in adults, type 1 diabetes, and type 2 diabetes. *Diabetes* **2008**, *57*, 1433–1437. [CrossRef]

51. Wellen, K.E.; Hotamisligil, G.S. Inflammation, stress, and diabetes. *J. Clin. Investig.* **2005**, *115*, 1111–1119. [CrossRef] [PubMed]

52. Tuomi, T.; Santoro, N.; Caprio, S.; Cai, M.; Weng, J.; Groop, L. The many faces of diabetes: A disease with increasing heterogeneity. *Lancet* **2014**, *383*, P1084–P1094. [CrossRef]

53. Hjort, R.; Ahlqvist, E.; Carlsson, P.O.; Grill, V.; Groop, L.; Martinell, M.; Rasouli, B.; Rosengren, A.; Tuomi, T.; Åsvold, B.O.; et al. Overweight, obesity and the risk of LADA: Results from a Swedish case–control study and the Norwegian HUNT study. *Diabetologia* **2018**, *61*, 1333–1343. [CrossRef] [PubMed]

54. Kharroubi, I.; Ladrière, L.; Cardozo, A.K.; Dogusan, Z.; Cnop, M.; Eizirik, D.L. Free fatty acids and cytokines induce pancreatic β-cell apoptosis by different mechanisms: Role of nuclear factor-κB and endoplasmic reticulum stress. *Endocrinology* **2004**, *145*, 5087–5096. [CrossRef] [PubMed]

55. Weitz, J.R.; Makhmutova, M.; Almaça, J.; Stertmann, J.; Aamodt, K.; Brissova, M.; Speier, S.; Rodriguez-Diaz, R.; Caicedo, A. Mouse pancreatic islet macrophages use locally released ATP to monitor beta cell activity. *Diabetologia* **2018**, *61*, 182–192. [CrossRef]

56. Cerf, M.E. Beta cell dysfunction and insulin resistance. *Front. Endocrinol.* **2013**, *4*, 37. [CrossRef]

57. Prentki, M.; Peyot, M.L.; Masiello, P.; Murthy Madiraju, S.R. Nutrient-induced metabolic stress, adaptation, detoxification, and toxicity in the pancreatic β-cell. *Diabetes* **2020**, *69*, 279–290. [CrossRef]

58. Schrimpe-Rutledge, A.C.; Fontès, G.; Gritsenko, M.A.; Norbeck, A.D.; Anderson, D.J.; Waters, K.M.; Adkins, J.N.; Smith, R.D.; Poitout, V.; Metz, T.O. Discovery of novel glucose-regulated proteins in isolated human pancreatic islets using LC-MS/MS-based proteomics. *J. Proteome Res.* **2012**, *11*, 3520–3532. [CrossRef]

59. Henquin, J.C.; Dufrane, D.; Nenquin, M. Nutrient control of insulin secretion in isolated normal human islets. *Diabetes* **2006**, *55*, 3470–3477. [CrossRef]

60. Schuit, F.; Flamez, D.; De Vos, A.; Pipeleers, D. Glucose-regulated gene expression maintaining the glucose-responsive state of β-cells. *Diabetes* **2002**, *51*, S326–S332. [CrossRef]

61. Oyadomari, S.; Araki, E.; Mori, M. Endoplasmic reticulum stress-mediated apoptosis in pancreatic β-cells. *Apoptosis* **2002**, *7*, 335–345. [CrossRef]

62. Schultheis, J.; Beckmann, D.; Mulac, D.; Müller, L.; Esselen, M.; Düfer, M. Nrf2 activation protects mouse beta cells from glucolipotoxicity by restoring mitochondrial function and physiological redox balance. *Oxid. Med. Cell. Longev.* **2019**, *2019*, 7518510. [CrossRef]

63. Hajmrle, C.; Smith, N.; Spigelman, A.F.; Dai, X.; Senior, L.; Bautista, A.; Ferdaoussi, M.; MacDonald, P.E. Interleukin-1 signaling contributes to acute islet compensation. *JCI Insight* **2016**, *1*, e86055. [CrossRef]

64. Henquin, J.C.; Ishiyama, N.; Nenquin, M.; Ravier, M.A.; Jonas, J.C. Signals and pools underlying biphasic insulin secretion. *Diabetes* **2002**, *5*, S60–S67. [CrossRef]

65. Chen, C.; Cohrs, C.M.; Stertmann, J.; Bozsak, R.; Speier, S. Human beta cell mass and function in diabetes: Recent advances in knowledge and technologies to understand disease pathogenesis. *Mol. Metab.* **2017**, *6*, 943–957. [CrossRef]

66. Cerf, M.E. High fat programming of beta cell compensation, exhaustion, death and dysfunction. *Pediatr. Diabetes* **2015**, *16*, 71–78. [CrossRef]

67. Mezza, T.; Muscogiuri, G.; Sorice, G.P.; Clemente, G.; Hu, J.; Pontecorvi, A.; Holst, J.J.; Giaccari, A.; Kulkarni, R.N. Insulin resistance alters islet morphology in nondiabetic humans. *Diabetes* **2014**, *63*, 994–1007. [CrossRef]

68. Yoneda, S.; Uno, S.; Iwahashi, H.; Fujita, Y.; Yoshikawa, A.; Kozawa, J.; Okita, K.; Takiuchi, D.; Eguchi, H.; Nagano, H.; et al. Predominance of β-cell neogenesis rather than replication in humans with an impaired glucose tolerance and newly diagnosed diabetes. *J. Clin. Endocrinol. Metab.* **2013**, *98*, 2053–2061. [CrossRef]

69. Burke, S.J.; Karlstad, M.D.; Regal, K.M.; Sparer, T.E.; Lu, D.; Elks, C.M.; Grant, R.W.; Stephens, J.M.; Burk, D.H.; Collier, J.J. CCL20 is elevated during obesity and differentially regulated by NF-κB subunits in pancreatic β-cells. *Biochim. Biophys. Acta Gene Regul. Mech.* **2015**, *1849*, 637–652. [CrossRef]

70. Burke, S.J.; Stadler, K.; Lu, D.; Gleason, E.; Han, A.; Donohoe, D.R.; Rogers, R.C.; Hermann, G.E.; Karlstad, M.D.; Collier, J.J. IL-1β reciprocally regulates chemokine and insulin secretion in pancreatic β-cells via NF-κB. *Am. J. Physiol. Endocrinol. Metab.* **2015**, *309*, E715–E726. [CrossRef]

71. Maedler, K.; Sergeev, P.; Ris, F.; Oberholzer, J.; Joller-Jemelka, H.I.; Spinas, G.A.; Kaiser, N.; Halban, P.A.; Donath, M.Y. Glucose-induced β cell production of IL-1β contributes to glucotoxicity in human pancreatic islets. *J. Clin. Investig.* **2002**, *110*, 851–860. [CrossRef]

72. Tersey, S.A.; Nishiki, Y.; Templin, A.T.; Cabrera, S.M.; Stull, N.D.; Colvin, S.C.; Evans-Molina, C.; Rickus, J.L.; Maier, B.; Mirmira, R.G. Islet β-cell endoplasmic reticulum stress precedes the onset of type 1 diabetes in the nonobese diabetic mouse model. *Diabetes* **2012**, *61*, 818–827. [CrossRef]

73. White, S.A.; Zhang, L.S.; Pasula, D.J.; Yang, Y.H.; Luciani, D.S. Bax and Bak jointly control survival and dampen the early unfolded protein response in pancreatic β-cells under glucolipotoxic stress. *Sci. Rep.* **2020**, *10*, 10986. [CrossRef]

74. Wei, M.C.; Zong, W.X.; Cheng, E.H.; Lindsten, T.; Panoutsakopoulou, V.; Ross, A.J.; Roth, K.A.; Macgregor, G.R.; Thompson, C.B.; Korsmeyer, S.J. Proapoptotic BAX and BAK: A requisite gateway to mitochondrial dysfunction and death. *Science* **2001**, *292*, 727–730. [CrossRef]

75. Scorrano, L.; Oakes, S.A.; Opferman, J.T.; Cheng, E.H.; Sorcinelli, M.D.; Pozzan, T.; Korsmeyer, S.J. BAX and BAK regulation of endoplasmic reticulum Ca^{2+}: A control point for apoptosis. *Science* **2003**, *300*, 135–139. [CrossRef]

76. Zong, W.X.; Li, C.; Hatzivassiliou, G.; Lindsten, T.; Yu, Q.C.; Yuan, J.; Thompson, C.B. Bax and Bak can localize to the endoplasmic reticulum to initiate apoptosis. *J. Cell Biol.* **2003**, *162*, 59–69. [CrossRef]

77. Gurzov, E.N.; Germano, C.M.; Cunha, D.A.; Ortis, F.; Vanderwinden, J.M.; Marchetti, P.; Zhang, L.; Eizirik, D.L. p53 up-regulated modulator of apoptosis (PUMA) activation contributes to pancreatic beta-cell apoptosis induced by proinflammatory cytokines and endoplasmic reticulum stress. *J. Biol. Chem.* **2010**, *285*, 19910–19920. [CrossRef]

78. Li, Y.; Ding, Y.; Zhang, Z.; Dai, X.; Jiang, Y.; Bao, L.; Li, Y. Grape seed proanthocyanidins ameliorate pancreatic beta-cell dysfunction and death in low-dose streptozotocin- and high-carbohydrate/high-fat diet-induced diabetic rats partially by regulating endoplasmic reticulum stress. *Nutr. Metab.* **2013**, *10*, 51. [CrossRef]

79. Donath, M.Y.; Shoelson, S.E. Type 2 diabetes as an inflammatory disease. *Nat. Rev. Immunol.* **2011**, *11*, 98–107. [CrossRef]

80. Novotny, G.W.; Lundh, M.; Backe, M.B.; Christensen, D.P.; Hansen, J.B.; Dahllöf, M.S.; Pallesen, E.M.; Mandrup-Poulsen, T. Transcriptional and translational regulation of cytokine signaling in inflammatory β-cell dysfunction and apoptosis. *Arch. Biochem. Biophys.* **2012**, *528*, 171–184. [CrossRef]

81. Eguchi, K.; Nagai, R. Islet inflammation in type 2 diabetes and physiology. *J. Clin. Investig.* **2017**, *127*, 14–23. [CrossRef] [PubMed]

82. Cerf, M.E. High fat programming of beta-cell failure. *Adv. Exp. Med. Biol.* **2010**, *654*, 77–89. [CrossRef] [PubMed]

Publisher's Note: MDPI stays neutral with regard to jurisdictional claims in published maps and institutional affiliations.

 *metabolites*

Review

Developmental Programming and Glucolipotoxicity: Insights on Beta Cell Inflammation and Diabetes

Marlon E. Cerf [1,2,3]

1   Grants, Innovation and Product Development, South African Medical Research Council,
    Tygerberg 7505, South Africa; marlon.cerf@mrc.ac.za
2   Biomedical Research and Innovation Platform, South African Medical Research Council,
    Tygerberg 7505, South Africa
3   Division of Medical Physiology, Department of Biomedical Sciences,
    Faculty of Medicine and Health Sciences, University of Stellenbosch, Tygerberg 7505, South Africa

Received: 14 August 2020; Accepted: 9 October 2020; Published: 4 November 2020

Abstract: Stimuli or insults during critical developmental transitions induce alterations in progeny anatomy, physiology, and metabolism that may be transient, sometimes reversible, but often durable, which defines programming. Glucolipotoxicity is the combined, synergistic, deleterious effect of simultaneously elevated glucose (chronic hyperglycemia) and saturated fatty acids (derived from high-fat diet overconsumption and subsequent metabolism) that are harmful to organs, micro-organs, and cells. Glucolipotoxicity induces beta cell death, dysfunction, and failure through endoplasmic reticulum and oxidative stress and inflammation. In beta cells, the misfolding of pro/insulin proteins beyond the cellular threshold triggers the unfolded protein response and endoplasmic reticulum stress. Consequentially there is incomplete and inadequate pro/insulin biosynthesis and impaired insulin secretion. Cellular stress triggers cellular inflammation, where immune cells migrate to, infiltrate, and amplify in beta cells, leading to beta cell inflammation. Endoplasmic reticulum stress reciprocally induces beta cell inflammation, whereas beta cell inflammation can self-activate and further exacerbate its inflammation. These metabolic sequelae reflect the vicious cycle of beta cell stress and inflammation in the pathophysiology of diabetes.

Keywords: beta cell death; beta cell dysfunction; beta cell failure; ER stress; hyperglycemia; obesity; oxidative stress; saturated fatty acids

1. Introduction

Beta cells are dynamic and respond to fluctuating demands for insulin. Inflammation contributes to the pathogenesis and is an underlying mechanism of several metabolic diseases. Developmental programming (hereafter programming) through a nutritional insult, such as maintenance on a high-fat diet (HFD) during fetal and early neonatal life, alters growth and developmental trajectories at the organ (e.g., pancreas), micro-organ (e.g., islets), and cellular (e.g., beta cell) levels that trigger the pathogenesis of metabolic diseases. In beta cells, high-fat programming (i.e., maintenance on a diet of ≥40% mainly saturated fat as energy during critical developmental windows) induces beta cell hypoplasia and hypotrophy (altered beta cell structure) that diminishes beta cell function (altered beta cell physiology) evident by impaired glucose-stimulated insulin secretion (GSIS) resulting in insufficient insulin release that results in and exacerbates hyperglycemia, as demonstrated in neonatal, weanling, and adult progeny [1–9]. High-fat programming also induces an altered metabolism of the substrates. For example, non-esterified fatty acids (NEFA) derived from fat metabolism have different profiles in circulation and organs that are dependent on the timing and duration of programming [10,11]. Chronic hyperglycemia contributes to further deterioration of beta cell function and to worsening insulin resistance, reflecting glucotoxicity. Chronic exposure to elevated circulating saturated fatty acids through high-fat programming induces lipotoxicity that similarly contributes to diminishing beta cell integrity and physiology and insulin resistance. Gluco- and lipo-toxicity typically co-exist as glucolipotoxicity. This article describes programming, glucolipotoxicity, and islet inflammation that precede and provoke beta cell inflammation and discusses their impact on beta cell physiology and dysfunction in the pathogenesis of diabetes.

2. Islet Inflammation

Macrophages are integral for inducing islet inflammation. In diabetes, intra-islet macrophage hyperplasia is the primary source of intra-islet proinflammatory cytokines [12]. Proinflammatory M1 macrophages produce and secrete interleukin 1 beta (IL1β), interleukin 6 (IL6), and tumor necrosis factor alpha (TNFα) to trigger inflammation [13], with IL1β central in islet inflammation, initiation and amplification [12]. In islets (in vitro and in vivo), M1 macrophages are the source of IL1β [12,14] and modulate beta cells' adaptive (i.e., compensatory) response to impaired function [15], characterized by beta cell dysfunction and failure. Physiologically, resident macrophages and cytokines maintain homeostasis in beta cell development and function [16]. However, metabolic diseases are often associated with chronic systemic inflammation [16] with islet and subsequent beta cell inflammation intrinsically linked to diabetes. In islet inflammation (insulitis), proinflammatory macrophage hyperplasia concomitant with elevated cytokine and chemokine concentrations contribute to impaired islet and beta cell function [16].

3. Programming and Islet Inflammation

Programming refers to a stimulus or insult during critical developmental transitions that induces alterations in offspring anatomy, physiology, and metabolism that may be transient or durable, and sometimes reversible. Nutrition, through a HFD, is one way to initiate programming. Pregnant HFD-fed C57/BL6J mice were obese with increased adiposity but not overtly diabetic [17]. However, a maternal HFD during gestation and lactation induced hepatic steatosis, adipose tissue inflammation, insulin resistance, and glucose intolerance [17]. In the islets of male progeny, there was increased oxidative stress concomitant with insulin resistance and worsening beta cell dysfunction after maintenance on a HFD from conception to weaning [17]. In pancreata from juvenile Japanese macaques (Macaca fuscata) maintained on a HFD during fetal life until to 13 months of age (high fat programmed primates), there was an increase in IL6 gene expression that correlated with a blunted first-phase insulin response reflecting early beta cell dysfunction [18]. In male progeny, there was increased pancreatic IL1β gene expression and fasting glucose concentrations [18]. Furthermore, in the juvenile primate pancreas, there was islet-associated macrophage hyperplasia concomitant with an increase in proinflammatory mediators that demonstrated that innate immune infiltration occurs prior to overt obesity or glucose dysregulation [18]. These metabolic derangements manifested prior to glucose dysregulation, revealing early events in the pathogenesis of diabetes.

In non-obese diabetic mice exposed to hyperglycemia in utero, the protective compensatory factor in response to islet stress, regenerating islet-derived protein 3 gamma (Reg3g), was decreased with a high fold change [19,20] that was deleterious for postnatal islet formation and/or maturation, thereby diminishing islet cell viability and function [21]. Furthermore, many upregulated genes were associated with pathways of inflammation and cell death [21]. As systemic inflammation was absent in progeny, the enhanced inflammation did not primarily induce islet dysfunction [21]. However, the increase in the inflammatory pathway enriched transcriptome in progeny exposed to hyperglycemia during late gestation was stimulated by greater islet cell susceptibility to death [21]. Programming with hyperglycemia therefore prompts beta cell stress, death, and inflammation. Thus, beta cell inflammation is a strong inducer of beta cell death, dysfunction, and failure, and is a predictor for developing diabetes.

Obesity is associated with immune cell hyperplasia [22–28] and inflammation. In humans and rodents with obese pregnancies, various cytokines and chemokines were elevated viz. IL1β, IL6,

IL10, TNF, interferon gamma (IFNγ) and monocyte chemoattractant protein 1 (MCP1/CCL2) [29–32], and obesity-associated maternal cytokines likely access the fetus via the placenta [33,34]. Further, maternal inflammation can initiate placental inflammation [35–38]; thus, the placenta is integral for conferring maternal obesity pathology to the fetus [39]. Therefore, obesity especially during pregnancy (maternal obesity) presents a major risk for progeny, as undesirable metabolic sequelae associated with obesity such as inflammation and hyperglycemia are conferred to progeny by their mothers.

This snapshot of the influence of programming through a HFD, hyperglycemia, and maternal obesity (often due to progeny exposed to a glucolipotoxic milieu during critical developmental phases) reveals the metabolic derangements that programming confers to progeny, e.g., hepatic steatosis, inflammation (adipose tissue, placenta and islets, and immune cell hyperplasia), insulin resistance, glucose intolerance, oxidative stress, and diminished islet and beta cell viability and function.

4. Glucolipotoxicity

4.1. Overview

Glucolipotoxicity is the combined deleterious consequences of elevated chronic glucose and saturated fatty acids (e.g., palmitic acid) concentrations on specific organs (e.g., the pancreas), and micro-organs (e.g., islets) and cells (e.g., beta cells) [40]. Glucolipotoxic-inducing diets with elevated glucose and saturated fatty acids prompt a persistent insulin demand, an increase in IAPP synthesis, potential increases in bacterial gut antigens, an increase in cytokine and chemokine production, and intra-islet inflammation that ultimately induces hyperglycemia and beta cell dysfunction [39]. With co-existing elevated lipidemia and glycemia, glucolipotoxicity manifests, resulting in metabolic alterations that drive the onset of diabetes [41–43].

However, during high-fat programming, glucolipotoxic effects can manifest earlier, given the particularly vulnerable life stage that the progeny are exposed to the insult. Hence, glucolipotoxicity induces derangements in beta cell structure and function, thereby inducing beta cell dysfunction; insulin resistance often follows as a consequence but can also contribute to beta cell dysfunction, and with progression to beta cell failure, diabetes ultimately manifests.

Beta cells initiate insulin transcription for pro/insulin biosynthesis and insulin exocytosis to restore glucose homeostasis, typically after ingesting a meal. Whereas glucose stimulates insulin transcription, pre-mRNA splicing and mRNA stability, proinsulin translation, maturation (to insulin) and insulin secretion, lipotoxicity and glucolipotoxicity impair several of these steps [44] and are associated with mitochondrial dysfunction [43,45]. However, persistent glucose excess, i.e., glucotoxicity, will also impair insulin processing and secretion. Ultimately, glucolipotoxicity induces beta cell dysfunction. Beta cell dysfunction, in relation to insulin secretion, presents as hyperinsulinemia during the adaptive or early pathogenesis of diabetes but evolves to hypoinsulinemia towards beta cell exhaustion and failure in the progression to overt diabetes.

Glucotoxicity is a key regulator of beta cells, whereas lipotoxicity and glucolipotoxicity were recently suggested to be less clinically relevant [46]. Elevated NEFA was acknowledged to be involved in diabetes by impairing insulin action, and despite higher NEFA concentrations in obese and diabetic individuals, there was reportedly no direct clinical evidence for beta cell lipotoxicity [46]. A supporting perspective proposed nutrient-induced metabolic stress as more relevant since NEFA are hard to trace in humans (in vivo), and therefore nutritional stress was suggested as more representative than glucolipotoxicity [47]. In a rat insulinoma beta cell line, treatment with palmitate resulted in ER expansion [48]. In db/db mouse islets and INS-1 cells treated with palmitic acid (to induce lipotoxicity), stimulator of interferon genes (STING), phosphorylated interferon regulatory factor 3 (IRF3), and IFNβ were upregulated, demonstrating that the STING-IRF3 pathway induces beta cell inflammation and apoptosis that contribute to beta cell dysfunction [49]. In another study, in human islets treated with palmitate and conducted in INS-1E cells using RNA sequencing and proteomics, there were 85 upregulated and 122 downregulated factors at the mRNA and protein levels implicated in oxidative stress, lipid metabolism, amino acid metabolism, and cell cycle pathways [50]. Lipotoxicity also affected several transcription factors implicated in metabolic and oxidative stress viz. liver X receptors (LXR), peroxisome proliferator-activated receptor alpha (PPARα), forkhead box protein O1 (FOXO1), and BTB and CNC homology 1, basic leucine zipper transcription factor 1 (BACH1) [50]. Different species and strains in animal models and varying experimental conditions in vitro and ex vivo have provided evidence for gluco-, lipo-, and glucolipotoxicity. For lipo- and glucolipotoxicity, the translational aspect to demonstrate clinical evidence and relevance remains a challenge. However, lipotoxicity and glucolipotoxicity remain relevant for beta cells, even potentially through indirect effects and under specific conditions, and hopefully, clinical studies will provide further evidence and demonstrate more direct lipotoxic and glucolipotoxic effects on beta cells to support their clinical relevance.

4.2. Islet and Beta Cell Glucolipotoxicity, Stress and Inflammation

Perpetual hyperglycemia and chronic saturated fatty acid exposure present as glucolipotoxicity, which contributes to islet and beta cell stress (endoplasmic reticulum (ER) and oxidative stress) and inflammation. Hyperglycemia and hyperlipidemia are traits of obesity and diabetes, and in diabetes, chronically high circulating NEFA concentrations, particularly saturated fatty acids, mediate a progressive decline in beta cell function leading to death, dysfunction [51] and failure.

In murine and human islets and beta cells, palmitate has been extensively used for the induction of ER and oxidative stress [41,52–54], whereas in rat islets, ex vivo exposure to elevated glucose, i.e., hyperglycemia, activated the unfolded protein response (UPR) [55]. This reflects the entwining of glucolipotoxicity, islet and beta cell stress. The exposure to high NEFA prompts superoxide (O_2K, an anion) and peroxynitrite (ONOOK) production by beta cell mitochondria and drives NOS_2 expression, which leads to nitric oxide (NO) production and induces ER and oxidative stress [51]. Palmitate co-cultured with healthy islets induced the release of the proinflammatory mediators (i.e., the cytokines IL6 and IL8, and the chemokine, chemokine (C-X-C motif) ligand 1 (CXCL1)) [53,56,57] with other proinflammatory cytokines viz. IL1β and TNFα also exhibited increased expression after co-culturing with palmitate [53]. Stearate and oleate (a monounsaturated fatty acid) also contribute to increased cytokine and chemokine expression [57], but inconsistently [53]. In islets, saturated fatty acids can induce chemokine (viz. Cxcl1 and Ccl3) gene expression [53,56] and initiate a microinflammatory response in the islets through the autocrine or paracrine effects of ILβ [12].

In isolated islets, the adverse effects of NEFA on GSIS are most profound with long-chain saturated fatty acids, e.g., palmitic acid [58]. Palmitic acid induces ER and oxidative stress, ceramide production, and jun *N*-terminal kinase (JNK) activation; all lead to inflammation [58,59]. In human islets, palmitic acid induced IL1β, IL6, IL8, TNFα, and chemokine (CCL2 and CXCL1) production, and also activated nuclear factor kappa-light-chain-enhancer of activated B cells (NFκB) [53]. These events demonstrate lipotoxic induced cellular stress and the provocation of inflammation in islets. Lipotoxicity and glucotoxicity can independently elicit their deleterious effects on islets and beta cells, e.g., inducing cellular stress and inflammation, also worsening diabetes [52,60]. However, they often converge as glucolipotoxicity, e.g., in obese and diabetic individuals with hyperglycemia and hyperlipidemia, which accelerates beta cell disintegration evident by cell death, dysfunction, and failure [42]. In murine islets, ethyl-palmitate infusion induced CCL2 and CXCL10 production through the toll-like receptor 4 (TLR4)-Myd88 pathway to facilitate M1 macrophage recruitment [61]. This establishes lipotoxicity as an aggravator of islet inflammation. In human diabetic pancreata and rodent diabetes, beta cells generated IL1β, i.e., were a source of hyperglycemic induced IL1β, and chronic hyperglycemia promoted intra-islet production of IL1β [62,63], which links glucotoxicity to islet inflammation. Further, IL1β alone, or IL1β with TNFα and IFNγ, prompts beta cell death, dysfunction [64–66], and failure with elevated glucose and NEFA concentrations, i.e., glucolipotoxicity, promoting the intra-islet production of IL1β [62,63]. Gluco-, lipo-, and glucolipotoxicity, therefore, induce islet and subsequently beta cell inflammation.

Glucolipotoxicity links beta cell stress to inflammation. Glucolipotoxicity alters mitochondrial function by impairing electron transport chain activity, which leads to increased reactive oxygen species (ROS) generation which, in turn, induces inflammation in beta cells and peripheral organs [13]. Mitochondria are close to the site of ROS generation and are therefore prone to oxidative stress [67]. The various signaling pathways that regulate mitochondrial ROS generation are involved in inflammation, ER and oxidative stress, mitochondrial biogenesis, energy demand, immune responses, and autophagy [68,69]. NEFA induce mtDNA damage dose-dependently and contribute to apoptosis [70]. Glucolipotoxicity diminishes beta cell mitochondria, inducing their fragmentation and rendering them incapable of fusion [71]. Sprague-Dawley rats maintained on a fetal and lactational HFD were insulin resistant with beta cell dysfunction (with enlarged insulin secretory granules) subsequent to reduced mtDNA content and altered mitochondrial gene expression [9]. Nutrient overload, such as supraphysiological glucose and NEFA concentrations, promotes ER stress, which impairs beta cells' secretory efficiency, and augments inflammation and oxidative stress. This further increases circulating glucose, i.e., hyperglycemia worsens and perpetuates [13].

5. ER and Oxidative Stress, Inflammation and Beta Cell Dysfunction

5.1. ER Stress

Cellular stress induced by ER and oxidative stress are integral to islet and beta cell inflammation, death, dysfunction, and failure. Beta cells are highly specialized for insulin biosynthesis, with pro/insulin accounting for 30–50% of total protein in beta cells [59]. ER stress is a major initiator of beta cell dysfunction. Physiologically, ~20% of proinsulin misfolds in beta cells, which is ~200,000 misfolded proinsulin molecules per minute [72]. Some ER protein misfolding is expected, with misfolding multiplying proportionally to protein complexity. However, ER stress manifests only when protein misfolding exceeds a threshold, which triggers the UPR to restore ER homeostasis [51]. Therefore, to maintain beta cell integrity and physiology, it is critical to ensure that pro/insulin misfolding does not exceed the threshold.

Beta cells often encounter ER overload due to increased metabolic demand, e.g., due to rapid and increased insulin biosynthesis and secretion in response to hyperglycemia [13] to maintain beta cell physiology. Healthy beta cells respond by increasing insulin biosynthesis >10 fold compared to un-/non-stimulated beta cells [73]. However, this may exceed the beta cells' folding capacity, thereby resulting in unfolded pro/insulin accumulation in the ER lumen, followed by ER stress [56,74] that provokes beta cell death, dysfunction, and ultimately, failure. ER stress can induce B cell lymphocyte (B cell) cytokine and chemokine gene expression [51] to initiate inflammation.

5.2. Oxidative Stress

The highly metabolically active beta cells are susceptible to oxidative stress due to low levels of some antioxidants, e.g., glutathione peroxidase (GPx) and catalase [75,76]. However, beta cells have other antioxidants, e.g., glutaredoxin and thioredoxin [75]. In beta cells, oxidative stress manifests due to increased ROS/reactive nitrogen species (RNS) production or the inefficiency of antioxidants to neutralize the increasing ROS/RNS levels [13].

In beta cells, ROS and RNS include free radicals, e.g., NO, O_2K and the hydroxyl radical (OH); non-radicals, e.g., hydrogen peroxide (H_2O_2); or anions, e.g., ONOOK [75,77,78]. Oxidative and ER stress induce inflammation due to prolonged UPR signaling with ROS/RNS activating NFκB and the inflammasome which drives B cell and IL1β secretion [79].

The close association of oxidative and ER stress is evident as oxidative stress induces protein misfolding through the disruption of the ER redox state and disulphide bond formation, and protein misfolding prompts ROS production [78]. The detrimental combination of beta cell dysfunction, induced by oxidative and ER stress, islet inflammation, beta cell death (by apoptosis) and impaired secretory pathway function, i.e., impaired GSIS, can therefore lead to the onset of diabetes [74,80,81].

5.3. Beta Cell Stress and Inflammation

Several mechanisms induce ER and oxidative stress that lead to beta cell inflammation, i.e., inflamed beta cells, and beta cell death, that then activates islet-resident cells, e.g., macrophages entwined with B, T, endothelial and dendritic cells, and myofibroblasts, concomitant with an increase in leukocyte recruitment [82–84]. These mechanisms are geared toward re-establishing proteostasis by restoring ER capacity for handling protein processing, folding, and trafficking [13]; but should proteostasis not be restored, beta cell death ensues [85].

In beta cells, IL1β (and other proinflammatory cytokines) induces ER stress and UPR activation, and promotes NFκB-dependent inducible nitric oxide synthase (iNOS) expression and NO production [13]. Cytokine and ER stress-induced beta cell death is associated as NO production induces ER calcium depletion and stress [86]. Oxidative stress contributes to ER stress and diminishes beta cell function such as impairing insulin transcription [87] and GSIS. In human islets, ER stress (via the RNA-dependent protein kinase (PKR)-like ER kinase (PERK) and inositol-requiring enzyme 1 (IRE1) pathways) induces thioredoxin-interacting protein (TXNIP), which in turn activates the nucleotide-binding domain leucine-rich repeat (NLR) and pyrin domain containing receptor 3 (NLRP3) inflammasome and IL1β production; thus, there is bidirectional ER stress and inflammation crosstalk [88]. Further, ER stress causes and is induced by chronic beta cell inflammation (i.e., both metabolic states cause and are caused by the other) due to their interconnectedness [13]. An increase in insulin biosynthesis (beyond the threshold) promotes ER overload, the UPR and ER stress, with persistent ER stress prompting apoptosis and IL1β release (through inflammasome activation) and local cytokines inducing NFκB activation that modulates the proapoptotic gene expression [13] which would lead to beta cell death. In beta cell dysfunction, intertwined oxidative and ER stress directly influences insulin biosynthesis, with ER stress-induced UPR [51] contributing to beta cell dysfunction and failure. This reflects the interrelations of ER and oxidative stress and inflammation; each process reciprocally contributes to and exacerbates the other processes.

6. Glucolipotoxicity: Interrelating Islet and Beta Cell Stress, Inflammation to Dysfunction

The islet cell production of proinflammatory mediators results in ER stress, oxidative stress, mitochondrial dysfunction, and islet cell dysfunction. Overnutrition results in obesity and increased adipose tissue mass, and with perpetual nutritional glucose and saturated fatty acid overload, leads! to ER stress. In humans, beta cell dedifferentiation in response to chronic overnutrition may be reversible within the first decade of diabetes [89]. Weight loss, through dietary intervention, restored the first-phase of insulin secretion, which was linked to reduced intrapancreatic triglyceride content [89]. ER stress, which results in the accumulation of misfolded pro/insulin, in turn, induces and increases beta cell inflammation that contributes to beta cell dysfunction. Some beta cell function is maintained but is compromised, and with time may lead to overt beta cell failure, where beta cells are non- or poorly functional. Beta cell dedifferentiation contributes to a reduced beta cell mass that would constrain functioning. Beta cell death is another consequence of ER and oxidative stress, and islet and beta cell inflammation.

Beta cell dysfunction initially presents in prediabetes (e.g., in obesity and insulin resistance) through their hypersecretion of insulin in response to elevated glucose concentrations. In obesity, hyperinsulinemia occurs due to the hypersecretion of insulin and compensatory response to insulin resistance [90]. In adolescents and adults, beta cell hypersecretion of insulin, in the absence of insulin resistance, was later linked to impaired glucose tolerance and diabetes [91]. This reflects the complexity of beta cell dysfunction and the progression of diabetes. Further, it provides clinical evidence that hyperinsulinemia may not be sustainable and eventually leads to the onset of diabetes as beta cells reach exhaustion. Individuals with diabetes risk factors such as familial diabetes and sedentary lifestyles had increased fat mass and dyslipidemia [91] and a worse prognosis.

In obesity and diabetes, there is an increase in circulating NEFA and ectopic fat accumulation in organs (e.g., in the liver and muscles due to adipose tissue capacitance being exceeded). In human islets treated with palmitate, there was increased islet triglyceride content concomitant with impaired function [92]. With chronic HFD feeding, there is an increase in systemic inflammation, and in IL6 and IL10 concentrations in the adipose tissue, resulting in an increase in adipose tissue inflammation. Adipose tissue insulin resistance, which is inflamed, induces modulated adipocyte sensory nerve secretion of human factors. This contributes to beta cell failure. These present inflammatory events, peripheral to the islet, along with other organ-specific inflammation, such as inflammation in the liver, muscle, heart, and kidneys that are induced by different diseases, but may be exacerbated by glucolipotoxicity. These prevailing inflammatory states, typical in obesity, co- and multi-morbidities, also interplay with islet inflammation. The interrelations of islet inflammation with systemic and organ-specific inflammation require further investigation.

Various programming insults such as fetal and lactational exposure to a high saturated fat diet, hyperglycemia or obesity (maternal obesity), can initiate glucolipotoxicity (Figure 1). Postnatal (post-lactational) exposure to these aggravators albeit less severe, when chronic (e.g., insulin resistance and glucose intolerance reflecting hyperglycemia, and long-term or severe obesity) also induce a glucolipotoxic milieu. These only represent some triggers of glucolipotoxicity due to excess glucose and NEFA that can be derived nutritionally (e.g., hyperlipidemia and HFD overconsumption) or by a compromised metabolic state (e.g., insulin resistance, glucose intolerance and obesity). An unfavorable early-life environment has durable effects on health and increases susceptibility to diabetes through epigenetic mechanisms [93,94]. Durable epigenetic modifications are programmed by maternal overnutrition—in C57BL6/J mice, hepatic insulin receptor substrate 2 (Irs2) and mitogen-activated protein kinase 4 (Map2k4) gene methylation occurred in progeny, thereby increasing their susceptibility to diabetes [7]. In human islets that were isolated and treated with IFNγ and IL1β, there was variable epigenetic remodeling [95] mediated by inflammatory transcription factor recruitment, DNA, looping and chromatin acetylation [96]. The upregulation of inflammatory and apoptotic factors [96] contributes to beta cell dysfunction. Glucolipotoxicity, induces beta cell death, dysfunction and failure through beta cell stress and inflammation (Figure 1).

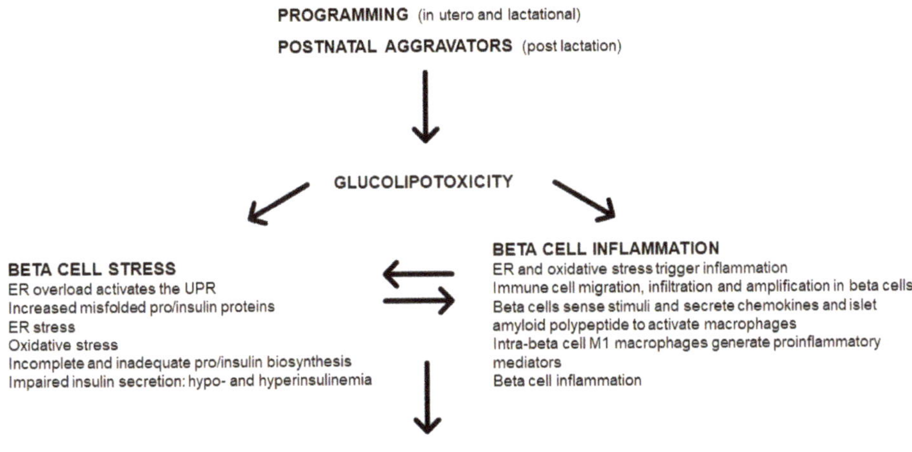

Figure 1. Programming, glucolipotoxicity, beta cell stress, and inflammation: convergence on beta cell death, dysfunction, and failure.

In overextended beta cells, glucolipotoxicity contributes to pro/insulin (and other) protein misfolding and accumulation in the ER lumen, resulting in ER overload which activates the UPR leading to ER stress (Figure 1). Further, in islets and beta cells, glucolipotoxicity can trigger and increase in ROS/RNS, leading to oxidative stress. ROS/RNS may regulate cell signaling pathways such as PI3K/Akt, mitogen-activated protein kinases (MAPK), JNK, and NFκB that govern cell proliferation, survival, and inflammation [13]. With resident islet and beta cell macrophages [56,97], inflammation can induce cell death through ROS/RNS [98], which links oxidative stress to islet and beta cell inflammation. ROS/RNS likely induces beta cell inflammation, beta cell death results in reduced GSIS, thereby inducing insulin resistance [99]. ER stress is also a source of ROS/RNS [99] and ER and oxidative stress are entwined and contribute to islet and beta cell inflammation. Consequentially there is incomplete and inadequate pro/insulin biosynthesis and impaired GSIS (Figure 1). Cellular stress triggers cellular inflammation, where immune cells migrate to, infiltrate and amplify in islet and beta cells leading to inflammation (Figure 1). Beta cells sense these stimuli and secrete IAPP to activate macrophages (Figure 1). The intra-islet M1 macrophages then generate more proinflammatory mediators thereby exacerbating beta cell inflammation. Beta cell stress and inflammation reciprocally cause and exacerbate each other, and over time lead to the onset of beta cell death, dysfunction, and failure (Figure 1).

Glucolipotoxicity induces beta cell stress with the increase in the UPR leading to ER and oxidative stress leading to increased beta cell inflammation (Figure 2). Further, glucolipotoxicity directly induces beta cell inflammation through an increase in proinflammatory mediators; within beta cells, macrophage hyperplasia and proinflammatory mediator production augment beta cell inflammation. This enhanced beta cell inflammation (through cellular stress or proinflammatory pathways) leads to hyperglycemia, reduced beta cell mass, beta cell death, dysfunction and failure, and ultimately, to diabetes (Figure 2). The entwined beta cell ER and oxidative stress and inflammation contribute to beta cell demise and failure leading to diabetes (Figure 2). These interrelations require further investigation to elucidate the mechanisms and sequence of events that lead to disease, and to identify novel agents that can counter their damaging effects on islet and beta cells. Further, systemically, obesity and inflammation can trigger macrophage migration, infiltration, and amplification, which contributes to reduced GSIS, and when prolonged, leads to diabetes (Figure 2).

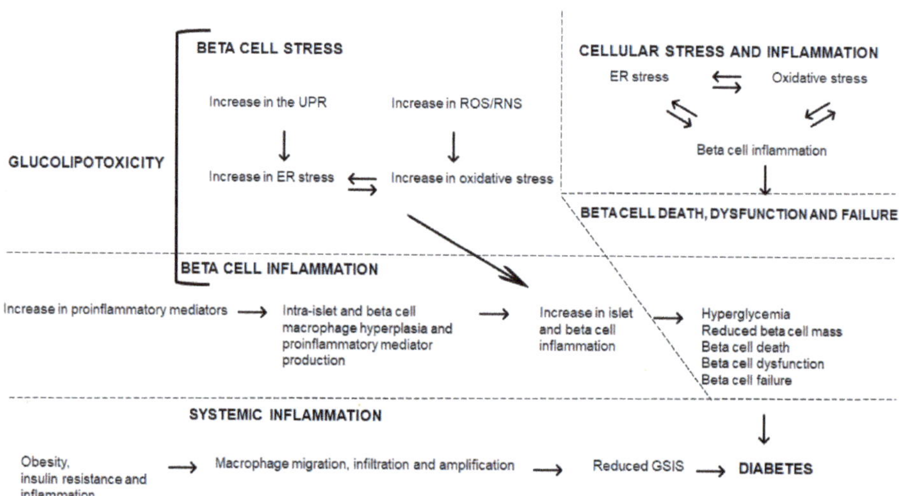

Figure 2. Glucolipotoxicity, beta cell stress, and inflammation: pathways to beta cell demise and diabetes.

Exacerbating obesity, adipose tissue insulin resistance and chronic overnutrition further prompt excess beta cell production of proinflammatory mediators. In pre-existing diabetes, islet and beta cell inflammation will exacerbate the disease and lead to further morbidities. ER stress induces and is induced by beta cell inflammation, whereas beta cell inflammation can self-activate and exacerbate beta cell inflammation. This reflects the vicious cycle of beta cell stress and inflammation in the pathophysiology of diabetes.

7. Conclusions

Programming, glucolipotoxicity, and the interactions of ER and oxidative stress in the pathogenesis and maintenance of disease require further unraveling and supporting evidence from clinical studies. Nutritionally and metabolically overloaded beta cells become stressed and inflamed with worsening outcomes for metabolic disease. Beta cell stress and inflammation require further investigation into adaptive mechanisms that evolve to mitigate cellular stress and inflammation, to identify strategies and targets for preserving beta cell physiology.

Funding: The author is supported by the National Research Foundation (South Africa).

Conflicts of Interest: The authors declare no conflict of interest.

References

1. Cerf, M.E.; Williams, K.; Nkomo, X.I.; Muller, C.J.; Du Toit, D.F.; Louw, J.; Wolfe-Coote, S.A. Islet cell response in the neonatal rat after exposure to a high-fat diet during pregnancy. *Am. J. Physiol. Regul. Integr. Comp. Physiol.* **2005**, *288*, R1122–R1128. [CrossRef] [PubMed]
2. Cerf, M.E.; Chapman, C.S.; Muller, C.J.; Louw, J. Gestational high-fat programming impairs insulin release and reduces Pdx-1 and glucokinase immunoreactivity in neonatal Wistar rats. *Metabolism* **2009**, *58*, 1787–1792. [CrossRef] [PubMed]
3. Cerf, M.E.; Muller, C.J.; Du Toit, D.F.; Louw, J.; Wolfe-Coote, S.A. Hyperglycaemia and reduced glucokinase expression in weanling offspring from dams maintained on a high-fat diet. *Br. J. Nutr.* **2006**, *95*, 391–396. [CrossRef] [PubMed]
4. Cerf, M.E.; Louw, J. Islet cell response to high fat programming in neonate, weanling and adolescent Wistar rats. *JOP* **2014**, *15*, 228–236. [CrossRef] [PubMed]
5. Gniuli, D.; Calcagno, A.; Caristo, M.E.; Mancuso, A.; Macchi, V.; Mingrone, G.; Vettor, R. Effects of high-fat diet exposure during fetal life on type 2 diabetes development in the progeny. *J. Lipid Res.* **2008**, *49*, 1936–1945. [CrossRef]
6. Karbaschi, R.; Zardooz, H.; Khodagholi, F.; Dargahi, L.; Salimi, M.; Rashidi, F. Maternal high-fat diet intensifies the metabolic response to stress in male rat offspring. *Nutr. Metab.* **2017**, *14*, 20. [CrossRef]
7. Zhang, Q.; Xiao, X.; Zheng, J.; Li, M.; Yu, M.; Ping, F.; Wang, T.; Wang, X. A maternal high-fat diet induces DNA methylation changes that contribute to glucose intolerance in offspring. *Front. Endocrinol.* **2019**, *10*, 871. [CrossRef]
8. Zambrano, E.; Sosa-Larios, T.; Calzada, L.; Ibáñez, C.A.; Mendoza-Rodríguez, C.A.; Morales, A.; Morimoto, S. Decreased basal insulin secretion from pancreatic islets of pups in a rat model of maternal obesity. *J. Endocrinol.* **2016**, *231*, 49–57. [CrossRef]
9. Taylor, P.D.; McConnell, J.; Khan, I.Y.; Holemans, K.; Lawrence, K.M.; Asare-Anane, H.; Persaud, S.J.; Jones, P.M.; Petrie, L.; Hanson, M.A.; et al. Impaired glucose homeostasis and mitochondrial abnormalities in offspring of rats fed a fat-rich diet in pregnancy. *Am. J. Physiol. Regul. Integr. Comp. Physiol.* **2005**, *288*, R134–R139. [CrossRef]
10. Cerf, M.E.; Louw, J.; Herrera, E. High fat diet exposure during fetal life enhances plasma and hepatic omega-6 fatty acid profiles in fetal Wistar rats. *Nutrients* **2015**, *7*, 7231–7241. [CrossRef]
11. Cerf, M.E.; Herrera, E. High fat diet administration during specific periods of pregnancy alters maternal fatty acid profiles in the near-term rat. *Nutrients* **2016**, *8*, 25. [CrossRef] [PubMed]
12. Eguchi, K.; Nagai, R. Islet inflammation in type 2 diabetes and physiology. *J. Clin. Investig.* **2017**, *127*, 14–23. [CrossRef] [PubMed]
13. Keane, K.N.; Cruzat, V.F.; Carlessi, R.; De Bittencourt, P.I.H.; Newsholme, P. Molecular events linking oxidative stress and inflammation to insulin resistance and β-cell dysfunction. *Oxid. Med. Cell. Longev.* **2015**, *29*, 457–474.e5. [CrossRef] [PubMed]

14. Masters, S.L.; Dunne, A.; Subramanian, S.L.; Hull, R.L.; Tannahill, G.M.; Sharp, F.A.; Becker, C.; Franchi, L.; Yoshihara, E.; Chen, Z.; et al. Activation of the NLRP3 inflammasome by islet amyloid polypeptide provides a mechanism for enhanced IL-1β in type 2 diabetes. *Nat. Immunol.* **2010**, *11*, 897–904. [CrossRef]

15. Ying, W.; Lee, Y.S.; Dong, Y.; Seidman, J.S.; Yang, M.; Isaac, R.; Seo, J.B.; Yang, B.H.; Wollam, J.; Riopel, M.; et al. Expansion of islet-resident macrophages leads to inflammation affecting β cell proliferation and function in obesity. *Cell Metab.* **2019**, *29*, 457–474.e5. [CrossRef]

16. Böni-Schnetzler, M.; Meier, D.T. Islet inflammation in type 2 diabetes. *Semin. Immunopathol.* **2019**, *41*, 501–513. [CrossRef]

17. Yokomizo, H.; Inoguchi, T.; Sonoda, N.; Sakaki, Y.; Maeda, Y.; Inoue, T.; Hirata, E.; Takei, R.; Ikeda, N.; Fujii, M.; et al. Maternal high-fat diet induces insulin resistance and deterioration of pancreatic β-cell function in adult offspring with sex differences in mice. *Am. J. Physiol. Endocrinol. Metab.* **2014**, *306*, E1163–E1175. [CrossRef]

18. Nicol, L.E.; Grant, W.F.; Comstock, S.M.; Nguyen, M.L.; Smith, M.S.; Grove, K.L.; Marks, D.L. Pancreatic inflammation and increased islet macrophages in insulin-resistant juvenile primates. *J. Endocrinol.* **2013**, *217*, 207–213. [CrossRef]

19. Marselli, L.; Thorne, J.; Dahiya, S.; Sgroi, D.C.; Sharma, A.; Bonner-Weir, S.; Marchetti, P.; Weir, G.C. Gene expression profiles of beta-cell enriched tissue obtained by laser capture microdissection from subjects with type 2 diabetes. *PLoS ONE* **2010**, *5*, e11499. [CrossRef]

20. Xia, F.; Cao, H.; Du, J.; Liu, X.; Liu, Y.; Xiang, M. Reg3g overexpression promotes β cell regeneration and induces immune tolerance in non-obese diabetic mouse model. *J. Leukoc. Biol.* **2016**, *99*, 1131–1140. [CrossRef]

21. Casasnovas, J.; Jo, Y.; Rao, X.; Xuei, X.; Brown, M.E.; Kua, K.L. High glucose alters fetal rat islet transcriptome and induces progeny islet dysfunction. *J. Endocrinol.* **2019**, *240*, 309–323. [CrossRef]

22. Bastard, J.P.; Maachi, M.; Lagathu, C.; Kim, M.J.; Caron, M.; Vidal, H.; Capeau, J.; Feve, B. Recent advances in the relationship between obesity, inflammation, and insulin resistance. *Eur. Cytokine Netw.* **2006**, *17*, 4–12.

23. Khaodhiar, L.; Ling, P.R.; Blackburn, G.L.; Bistrian, B.R. Serum levels of interleukin-6 and C-reactive protein correlate with body mass index across the broad range of obesity. *J. Parenter. Enter. Nutr.* **2004**, *28*, 410–415. [CrossRef] [PubMed]

24. Sartipy, P.; Loskutoff, D.J. Monocyte chemoattractant protein 1 in obesity and insulin resistance. *Proc. Natl. Acad. Sci. USA* **2003**, *100*, 7265–7270. [CrossRef] [PubMed]

25. Weisberg, S.P.; McCann, D.; Desai, M.; Rosenbaum, M.; Leibel, R.L.; Ferrante, A.W. Obesity is associated with macrophage accumulation in adipose tissue. *J. Clin. Investig.* **2003**, *112*, 1796–1808. [CrossRef] [PubMed]

26. Das, U.N. Is obesity an inflammatory condition? *Nutrition* **2001**, *17*, 953–966. [CrossRef]

27. Fried, S.K.; Bunkin, D.A.; Greenberg, A.S. Omental and subcutaneous adipose tissues of obese subjects release interleukin-6: Depot difference and regulation by glucocorticoid. *J. Clin. Endocrinol. Metab.* **1998**, *83*, 847–850. [CrossRef]

28. Hotamisligil, G.S.; Shargill, N.S.; Spiegelman, B.M. Adipose expression of tumor necrosis factor-alpha: Direct role in obesity-linked insulin resistance. *Science* **1993**, *259*, 87–91. [CrossRef]

29. Catalano, P.M.; Presley, L.; Minium, J.; Hauguel-de Mouzon, S. Fetuses of obese mothers develop insulin resistance in utero. *Diabetes Care* **2009**, *32*, 1076–1080. [CrossRef]

30. Madan, J.C.; Davis, J.M.; Craig, W.Y.; Collins, M.; Allan, W.; Quinn, R.; Dammann, O. Maternal obesity and markers of inflammation in pregnancy. *Cytokine* **2009**, *47*, 61–64. [CrossRef]

31. Roberts, K.A.; Riley, S.C.; Reynolds, R.M.; Barr, S.; Evans, M.; Statham, A.; Hor, K.; Jabbour, H.N.; Norman, J.E.; Denison, F.C. Placental structure and inflammation in pregnancies associated with obesity. *Placenta* **2011**, *32*, 247–254. [CrossRef]

32. Keępczyńska, M.A.; Wargent, E.T.; Cawthorne, M.A.; Arch, J.R.S.; O'Dowd, J.F.; Stocker, C.J. Circulating levels of the cytokines IL10, IFNγ and resistin in an obese mouse model of developmental programming. *J. Dev. Orig. Health Dis.* **2013**, *4*, 491–498. [CrossRef] [PubMed]

33. Zaretsky, M.V.; Alexander, J.M.; Byrd, W.; Bawdon, R.E. Transfer of inflammatory cytokines across the placenta. *Obstet. Gynecol.* **2004**, *103*, 546–550. [CrossRef]

34. Dahlgren, J.; Samuelsson, A.M.; Jansson, T.; Holmäng, A. Interleukin-6 in the maternal circulation reaches the rat fetus in mid-gestation. *Pediatr. Res.* **2006**, *60*, 147–151. [CrossRef]

35. Zhu, M.J.; Du, M.; Nathanielsz, P.W.; Ford, S.P. Maternal obesity up-regulates inflammatory signaling pathways and enhances cytokine expression in the mid-gestation sheep placenta. *Placenta* **2010**, *31*, 387–391. [CrossRef]

36. Challier, J.C.; Basu, S.; Bintein, T.; Minium, J.; Hotmire, K.; Catalano, P.M.; Hauguel-de Mouzon, S. Obesity in pregnancy stimulates macrophage accumulation and inflammation in the placenta. *Placenta* **2008**, *29*, 274–281. [CrossRef]

37. Hauguel-de Mouzon, S.; Guerre-Millo, M. The placenta cytokine network and inflammatory signals. *Placenta* **2006**, *27*, 794–798. [CrossRef]

38. Aaltonen, R.; Heikkinen, T.; Hakala, K.; Laine, K.; Alanen, A. Transfer of proinflammatory cytokines across term placenta. *Obstet. Gynecol.* **2005**, *106*, 802–807. [CrossRef]

39. Kim, D.W.; Young, S.L.; Grattan, D.R.; Jasoni, C.L. Obesity during pregnancy disrupts placental morphology, cell proliferation, and inflammation in a sex-specific manner across gestation in the mouse. *Biol. Reprod.* **2014**, *90*, 130. [CrossRef]

40. Poitout, V.; Amyot, J.; Semache, M.; Zarrouki, B.; Hagman, D.; Fontés, G. Glucolipotoxicity of the pancreatic beta cell. *Biochim. Biophys. Acta Mol. Cell Biol. Lipids* **2010**, *1801*, 289–298. [CrossRef]

41. Cunha, D.A.; Hekerman, P.; Ladrière, L.; Bazarra-Castro, A.; Ortis, F.; Wakeham, M.C.; Moore, F.; Rasschaert, J.; Cardozo, A.K.; Bellomo, E.; et al. Initiation and execution of lipotoxic ER stress in pancreatic β-cells. *J. Cell Sci.* **2008**, *121*, 2308–2318. [CrossRef] [PubMed]

42. El-Assaad, W.; Buteau, J.; Peyot, M.L.; Nolan, C.; Roduit, R.; Hardy, S.; Joly, E.; Dbaibo, G.; Rosenberg, L.; Prentki, M. Saturated fatty acids synergize with elevated glucose to cause pancreatic β-cell death. *Endocrinology* **2003**, *144*, 4154–4163. [CrossRef] [PubMed]

43. Somesh, B.P.; Verma, M.K.; Sadasivuni, M.K.; Mammen-Oommen, A.; Biswas, S.; Shilpa, P.C.; Reddy, A.K.; Yateesh, A.N.; Pallavi, P.M.; Nethra, S.; et al. Chronic glucolipotoxic conditions in pancreatic islets impair insulin secretion due to dysregulated calcium dynamics, glucose responsiveness and mitochondrial activity. *BMC Cell Biol.* **2013**, *14*, 31. [CrossRef] [PubMed]

44. Lytrivi, M.; Castell, A.L.; Poitout, V.; Cnop, M. Recent insights into mechanisms of β-cell lipo- and glucolipotoxicity in type 2 diabetes. *J. Mol. Biol.* **2020**, *432*, 1514–1534. [CrossRef]

45. Barlow, J.; Affourtit, C. Novel insights into pancreatic β-cell glucolipotoxicity from real-time functional analysis of mitochondrial energy metabolism in INS-1E insulinoma cells. *Biochem. J.* **2013**, *456*, 417–426. [CrossRef]

46. Weir, G.C. Glucolipotoxicity, β-cells, and diabetes: The emperor has no clothes. *Diabetes* **2020**, *69*, 273–278. [CrossRef]

47. Prentki, M.; Peyot, M.L.; Masiello, P.; Murthy Madiraju, S.R. Nutrient-induced metabolic stress, adaptation, detoxification, and toxicity in the pancreatic β-cell. *Diabetes* **2020**, *69*, 279–290. [CrossRef]

48. Pinnick, K.; Neville, M.; Clark, A.; Fielding, B. Reversibility of metabolic and morphological changes associated with chronic exposure of pancreatic islet β-cells to fatty acids. *J. Cell. Biochem.* **2010**, *109*, 683–692. [CrossRef]

49. Hu, H.Q.; Qiao, J.T.; Liu, F.Q.; Wang, J.B.; Sha, S.; He, Q.; Cui, C.; Song, J.; Zang, N.; Wang, L.S.; et al. The STING-IRF3 pathway is involved in lipotoxic injury of pancreatic β cells in type 2 diabetes. *Mol. Cell. Endocrinol.* **2020**, 11089. [CrossRef]

50. Lytrivi, M.; Ghaddar, K.; Lopes, M.; Rosengren, V.; Piron, A.; Yi, X.; Johansson, H.; Lehtiö, J.; Igoillo-Esteve, M.; Cunha, D.A.; et al. Combined transcriptome and proteome profiling of the pancreatic β-cell response to palmitate unveils key pathways of β-cell lipotoxicity. *BMC Genom.* **2020**, *21*, 590. [CrossRef]

51. Hasnain, S.Z.; Prins, J.B.; McGuckin, M.A. Oxidative and endoplasmic reticulum stress in β-cell dysfunction in diabetes. *J. Mol. Endocrinol.* **2016**, *56*, R33–R54. [CrossRef] [PubMed]

52. Cnop, M.; Ladrière, L.; Igoillo-Esteve, M.; Moura, R.F.; Cunha, D.A. Causes and cures for endoplasmic reticulum stress in lipotoxic β-cell dysfunction. *Diabetes Obes. Metab.* **2010**, *12*, 76–82. [CrossRef] [PubMed]

53. Igoillo-Esteve, M.; Marselli, L.; Cunha, D.A.; Ladrière, L.; Ortis, F.; Grieco, F.A.; Dotta, F.; Weir, G.C.; Marchetti, P.; Eizirik, D.L.; et al. Palmitate induces a pro-inflammatory response in human pancreatic islets that mimics CCL2 expression by beta cells in type 2 diabetes. *Diabetologia* **2010**, *53*, 1395–1405. [CrossRef] [PubMed]

54. Hasnain, S.Z.; Borg, D.J.; Harcourt, B.E.; Tong, H.; Sheng, Y.H.; Ng, C.P.; Das, I.; Wang, R.; Chen, A.C.; Loudovaris, T.; et al. Glycemic control in diabetes is restored by therapeutic manipulation of cytokines that regulate beta cell stress. *Nat. Med.* **2014**, *20*, 1417–1426. [CrossRef]

55. Elouil, H.; Bensellam, M.; Guiot, Y.; Vander Mierde, D.; Pascal, S.M.; Schuit, F.C.; Jonas, J.C. Acute nutrient regulation of the unfolded protein response and integrated stress response in cultured rat pancreatic islets. *Diabetologia* **2007**, *50*, 1442–1452. [CrossRef]

56. Ehses, J.A.; Perren, A.; Eppler, E.; Ribaux, P.; Pospisilik, J.A.; Maor-Cahn, R.; Gueripel, X.; Ellingsgaard, H.; Schneider, M.K.; Biollaz, G.; et al. Increased number of islet-associated macrophages in type 2 diabetes. *Diabetes* **2007**, *56*, 2356–2370. [CrossRef]

57. Böni-Schnetzler, M.; Boller, S.; Debray, S.; Bouzakri, K.; Meier, D.T.; Prazak, R.; Kerr-Conte, J.; Pattou, F.; Ehses, J.A.; Schuit, F.C.; et al. Free fatty acids induce a proinflammatory response in islets via the abundantly expressed interleukin-1 receptor I. *Endocrinology* **2009**, *150*, 5218–5229. [CrossRef]

58. Van Raalte, D.H.; Diamant, M. Glucolipotoxicity and beta cells in type 2 diabetes mellitus: Target for durable therapy? *Diabetes Res. Clin. Pract.* **2011**, *93*, S37–S46. [CrossRef]

59. Imai, Y.; Dobrian, A.D.; Morris, M.A.; Nadler, J.L. Islet inflammation: A unifying target for diabetes treatment? *Trends Endocrinol. Metab.* **2013**, *24*, 351–360. [CrossRef]

60. Donath, M.Y.; Gross, D.J.; Cerasi, E.; Kaiser, N. Hyperglycemia-induced beta-cell apoptosis in pancreatic islets of Psammomys obesus during development of diabetes. *Diabetes* **1999**, *48*, 738–744. [CrossRef]

61. Eguchi, K.; Manabe, I.; Oishi-Tanaka, Y.; Ohsugi, M.; Kono, N.; Ogata, F.; Yagi, N.; Ohto, U.; Kimoto, M.; Miyake, K.; et al. Saturated fatty acid and TLR signaling link β cell dysfunction and islet inflammation. *Cell Metab.* **2012**, *15*, 518–533. [CrossRef]

62. Ehses, J.A.; Böni-Schnetzler, M.; Faulenbach, M.; Donath, M.Y. Macrophages, cytokines and β-cell death in type 2 diabetes. *Biochem. Soc. Trans.* **2008**, *36*, 340–342. [CrossRef] [PubMed]

63. Maedler, K.; Sergeev, P.; Ris, F.; Oberholzer, J.; Joller-Jemelka, H.I.; Spinas, G.A.; Kaiser, N.; Halban, P.A.; Donath, M.Y. Glucose-induced β cell production of IL-1β contributes to glucotoxicity in human pancreatic islets. *J. Clin. Investig.* **2002**, *110*, 851–860. [CrossRef]

64. Mandrup-Poulsen, T. The role of interleukin-1 in the pathogenesis of IDDM. *Diabetologia* **1996**, *39*, 1005–1029. [CrossRef] [PubMed]

65. Rabinovitch, A.; Suarez-Pinzon, W.L.; Shi, Y.; Morgan, A.R.; Bleackley, R.C. DNA fragmentation is an early event in cytokine-induced islet beta-cell destruction. *Diabetologia* **1994**, *37*, 733–738. [CrossRef]

66. Iwahashi, H.; Hanafusa, T.; Eguchi, Y.; Nakajima, H.; Miyagawa, J.; Itoh, N.; Tomita, K.; Namba, M.; Kuwajima, M.; Noguchi, T.; et al. Cytokine-induced apoptotic cell death in a mouse pancreatic beta-cell line: Inhibition by Bcl-2. *Diabetologia* **1996**, *39*, 530–536. [CrossRef] [PubMed]

67. Li, N.; Frigerio, F.; Maechler, P. The sensitivity of pancreatic β-cells to mitochondrial injuries triggered by lipotoxicity and oxidative stress. *Biochem. Soc. Trans.* **2008**, *36*, 930–934. [CrossRef] [PubMed]

68. Sena, L.A.; Chandel, N.S. Physiological roles of mitochondrial reactive oxygen species. *Mol. Cell* **2012**, *48*, 158–167. [CrossRef]

69. Bolisetty, S.; Jaimes, E.A. Mitochondria and reactive oxygen species: Physiology and pathophysiology. *Int. J. Mol. Sci.* **2013**, *14*, 6306–6344. [CrossRef]

70. Grishko, V.; Rachek, L.; Musiyenko, S.; LeDoux, S.P.; Wilson, G.L. Involvement of mtDNA damage in free fatty acid-induced apoptosis. *Free Radic. Biol. Med.* **2005**, *38*, 755–762. [CrossRef]

71. Molina, A.J.; Wikstrom, J.D.; Stiles, L.; Las, G.; Mohamed, H.; Elorza, A.; Walzer, G.; Twig, G.; Katz, S.; Corkey, B.E.; et al. Mitochondrial networking protects β-cells from nutrient-induced apoptosis. *Diabetes* **2009**, *58*, 2303–2315. [CrossRef] [PubMed]

72. Sun, J.; Cui, J.; He, Q.; Chen, Z.; Arvan, P.; Liu, M. Proinsulin misfolding and endoplasmic reticulum stress during the development and progression of diabetes. *Mol. Asp. Med.* **2015**, *42*, 105–118. [CrossRef] [PubMed]

73. Eizirik, D.L.; Cardozo, A.K.; Cnop, M. The role for endoplasmic reticulum stress in diabetes mellitus. *Endocr. Rev.* **2008**, *29*, 42–61. [CrossRef] [PubMed]

74. Eizirik, D.L.; Miani, M.; Cardozo, A.K. Signalling danger: Endoplasmic reticulum stress and the unfolded protein response in pancreatic islet inflammation. *Diabetologia* **2013**, *56*, 234–241. [CrossRef] [PubMed]

75. Newsholme, P.; Rebelato, E.; Abdulkader, F.; Krause, M.; Carpinelli, A.; Curi, R. Reactive oxygen and nitrogen species generation, antioxidant defenses, and β-cell function: A critical role for amino acids. *J. Endocrinol.* **2012**, *214*, 11–20. [CrossRef]

76. Gehrmann, W.; Elsner, M.; Lenzen, S. Role of metabolically generated reactive oxygen species for lipotoxicity in pancreatic β-cells. *Diabetes Obes. Metab.* **2010**, *12*, 149–158. [CrossRef]

77. Lenzen, S. Oxidative stress: The vulnerable β-cell. *Biochem. Soc. Trans.* **2008**, *36*, 343–347. [CrossRef]

78. Cao, S.S.; Kaufman, R.J. Endoplasmic reticulum stress and oxidative stress in cell fate decision and human disease. *Antioxid. Redox Signal.* **2014**, *21*, 396–413. [CrossRef] [PubMed]

79. Menu, P.; Mayor, A.; Zhou, R.; Tardivel, A.; Ichijo, H.; Mori, K.; Tschopp, J. ER stress activates the NLRP3 inflammasome via an UPR-independent pathway. *Cell Death Dis.* **2012**, *3*, e261. [CrossRef]

80. Donath, M.Y.; Dalmas, É.; Sauter, N.S.; Böni-Schnetzler, M. Inflammation in obesity and diabetes: Islet dysfunction and therapeutic opportunity. *Cell Metab.* **2013**, *17*, 860–872. [CrossRef]

81. Montane, J.; Cadavez, L.; Novials, A. Stress and the inflammatory process: A major cause of pancreatic cell death in type 2 diabetes. *Diabetes Metab. Syndr. Obes. Targets Ther.* **2014**, *7*, 25–34. [CrossRef]

82. Toyama, H.; Takada, M.; Tanaka, T.; Suzuki, Y.; Kuroda, Y. Characterization of islet-infiltrating immunocytes after pancreas preservation by two-layer (UW/perfluorochemical) cold storage method. *Transpl. Proc.* **2003**, *35*, 1503–1505. [CrossRef]

83. Coppieters, K.T.; Dotta, F.; Amirian, N.; Campbell, P.D.; Kay, T.W.; Atkinson, M.A.; Roep, B.O.; von Herrath, M.G. Demonstration of islet-autoreactive CD8 T cells in insulitic lesions from recent onset and long-term type 1 diabetes patients. *J. Exp. Med.* **2012**, *209*, 51–60. [CrossRef] [PubMed]

84. Calderon, B.; Carrero, J.A.; Ferris, S.T.; Sojka, D.K.; Moore, L.; Epelman, S.; Murphy, K.M.; Yokoyama, W.M.; Randolph, G.J.; Unanue, E.R. The pancreas anatomy conditions the origin and properties of resident macrophages. *J. Exp. Med.* **2015**, *212*, 1497–1512. [CrossRef]

85. Xu, C.; Bailly-Maitre, B.; Reed, J.C. Endoplasmic reticulum stress: Cell life and death decisions. *J. Clin. Invest.* **2005**, *115*, 2656–2664. [CrossRef]

86. Oyadomari, S.; Takeda, K.; Takiguchi, M.; Gotoh, T.; Matsumoto, M.; Wada, I.; Akira, S.; Araki, E.; Mori, M. Nitric oxide-induced apoptosis in pancreatic β cells is mediated by the endoplasmic reticulum stress pathway. *Proc. Natl. Acad. Sci. USA* **2001**, *98*, 10845–10850. [CrossRef]

87. Kaneto, H.; Matsuoka, T.A. Role of pancreatic transcription factors in maintenance of mature β-cell function. *Int. J. Mol. Sci.* **2015**, *16*, 6281–6297. [CrossRef]

88. Oslowski, C.M.; Hara, T.; O'Sullivan-Murphy, B.; Kanekura, K.; Lu, S.; Hara, M.; Ishigaki, S.; Zhu, L.J.; Hayashi, E.; Hui, S.T.; et al. Thioredoxin-interacting protein mediates ER stress-induced β cell death through initiation of the inflammasome. *Cell Metab.* **2012**, *16*, 265–273. [CrossRef]

89. White, M.G.; Shaw, J.A.; Taylor, R. Type 2 diabetes: The pathologic basis of reversible β-cell dysfunction. *Diabetes Care* **2016**, *39*, 2080–2088. [CrossRef]

90. Ferrannini, E.; Natali, A.; Bell, P.; Cavallo-Perin, P.; Lalic, N.; Mingrone, G. Insulin resistance and hypersecretion in obesity. *J. Clin. Invest.* **1997**, *100*, 1166–1173. [CrossRef]

91. Tricò, D.; Natali, A.; Arslanian, S.; Mari, A.; Ferrannini, E. Identification, pathophysiology, and clinical implications of primary insulin hypersecretion in nondiabetic adults and adolescents. *JCI Insight* **2018**, *3*, e124912. [CrossRef] [PubMed]

92. Lalloyer, F.; Vandewalle, B.; Percevault, F.; Torpier, G.; Kerr-Conte, J.; Oosterveer, M.; Paumelle, R.; Fruchart, J.C.; Kuipers, F.; Pattou, F.; et al. Peroxisome proliferator-activated receptor α improves pancreatic adaptation to insulin resistance in obese mice and reduces lipotoxicity in human islets. *Diabetes* **2006**, *55*, 1605–1613. [CrossRef] [PubMed]

93. Aleliunas, R.E.; Aljaadi, A.M.; Laher, I.; Glier, M.B.; Green, T.J.; Murphy, M.; Miller, J.W.; Devlin, A.M. Folic acid supplementation of female mice, with or without vitamin B-12, before and during pregnancy and lactation programs adiposity and vascular health in adult male offspring. *J. Nutr.* **2016**, *146*, 688–696. [CrossRef] [PubMed]

94. Reusens, B.; Theys, N.; Dumortier, O.; Goosse, K.; Remacle, C. Maternal malnutrition programs the endocrine pancreas in progeny. *Am. J. Clin. Nutr.* **2011**, *94*, 1824S–1829S. [CrossRef]

95. Peters, L.; Posgai, A.; Brusko, T.M. Islet–immune interactions in type 1 diabetes: The nexus of beta cell destruction. *Clin. Exp. Immunol.* **2019**, *198*, 326–340. [CrossRef]

96. Ramos-Rodríguez, M.; Raurell-Vila, H.; Colli, M.L.; Alvelos, M.I.; Subirana-Granés, M.; Juan-Mateu, J.; Norris, R.; Turatsinze, J.V.; Nakayasu, E.S.; Webb-Robertson, B.J.; et al. The impact of proinflammatory cytokines on the β-cell regulatory landscape provides insights into the genetics of type 1 diabetes. *Nat. Genet.* **2019**, *51*, 1588–1595. [CrossRef]

97. Böni-Schnetzler, M.; Ehses, J.A.; Faulenbach, M.; Donath, M.Y. Insulitis in type 2 diabetes. *Diabetes Obes. Metab.* **2008**, *10*, 201–204. [CrossRef]

98. Choudhury, S.; Ghosh, S.; Gupta, P.; Mukherjee, S.; Chattopadhyay, S. Inflammation-induced ROS generation causes pancreatic cell death through modulation of Nrf2/NF-κB and SAPK/JNK pathway. *Free Radic. Res.* **2015**, *49*, 1371–1383. [CrossRef]

99. Burgos-Morón, E.; Abad-Jiménez, Z.; Martinez de Marañon, A.; Iannantuoni, F.; Escribano-López, I.; López-Domènech, S.; Salom, C.; Jover, A.; Mora, V.; Roldan, I.; et al. Relationship between oxidative stress, ER stress, and inflammation in type 2 diabetes: The battle continues. *J. Clin. Med.* **2019**, *8*, 1385. [CrossRef]

Publisher's Note: MDPI stays neutral with regard to jurisdictional claims in published maps and institutional affiliations.

MDPI

St. Alban-Anlage 66

4052 Basel

Switzerland

Tel. +41 61 683 77 34

Fax +41 61 302 89 18

www.mdpi.com

Metabolites Editorial Office

E-mail: metabolites@mdpi.com

www.mdpi.com/journal/metabolites